A SURVEY OF NONLINEAR DYNAMICS

("Chaos Theory")

A SURVEY OF NONLINEAR DYNAMICS

("*Chaos Theory*")

R. L. Ingraham
Department of Physics
New Mexico State University
USA

World Scientific
Singapore • New Jersey • London • Hong Kong

Published by

World Scientific Publishing Co. Pte. Ltd.
P O Box 128, Farrer Road, Singapore 9128
USA office: Suite 1B, 1060 Main Street, River Edge, NJ 07661
UK office: 73 Lynton Mead, Totteridge, London N20 8DH

Library of Congress Cataloging-in-Publication Data

Ingraham, R. L. (Richard Lee), 1923–
A survey of nonlinear dynamics : "chaos theory" / R.L. Ingraham.
p. cm.
Includes bibliographical references and index.
1. Chaotic behavior in systems. I. Title.
Q172.5.C45154 1991
003'.7--dc20 91-41291
CIP

ISBN 981-02-0777-8

Printed in Singapore by JBW Printers & Binders Pte. Ltd.

PREFACE

This report is intended to give a survey of the whole field of nonlinear dynamics (or "chaos theory," as it is popularly called) in a compressed form. It is slightly expanded from a series of lectures given over the space of a single month in 1989. This young and rapidly growing field is already very extensive, so that this survey cannot be deep or detailed. In particular, no pretense of mathematical rigor is made. But I do insist on stating key definitions or theorems carefully so that the reader need not settle for just a qualitative, intuitive understanding. My intention is to touch on the main ideas so that the reader can see if his or her special discipline fits in anywhere and if so, can get an approximate notion of what new ideas or possibilities nonlinear dynamics brings to that field. The cited literature then allows the reader to proceed further if he or she desires.

I thank Harry Auvermann for suggesting that I give these lectures in the first place, and David Bandelier for a beautiful job of turning my manuscript into print. Thanks to David Tofsted for invaluable aid in reformatting the report into book form. I acknowledge support from the Atmospheric Sciences Laboratory (representative Harry Auvermann) under the auspices of the Scientific Services Program administered by Battelle for the Army Research Office (D.O. 2008, Contract DAAL03-86-D-0001).

CONTENTS

A SURVEY OF NONLINEAR DYNAMICS

("*Chaos Theory*")

1. INTRODUCTORY REMARKS

1.1 Linear Versus Nonlinear

A *dynamics* describes the time evolution of a system. As such, the concept is not confined to physics, but occurs in many other fields as well – in related sciences like engineering, chemistry, and biology, but also in ecology, economics, etc. A *nonlinear* dynamics describes the time-evolution via nonlinear equations of motion, which may be ordinary differential equations, partial differential equations, difference equations, iteration of maps, etc. Nonlinear motion equations have been around a long time – since the beginning of science, in fact – so why the sudden blooming of nonlinear dynamics as a new discipline in the last 20 years or so?

The answer is that up to that time nonlinear equations were regarded as not essentially different from linear ones – more complicated and difficult to solve, of course, but nothing that suitably refined linear approximations couldn't handle. Analytic ("closed form") solutions were emphasized in textbooks with the confident expectation that "nonanalytic" solutions, if they existed, formed a small subset of all solutions which didn't greatly add to the understanding of the phenomena. But about 20 years ago it was realized that nonlinear equations are essentially different from linear ones, that they possess properties which can never be captured by linear approximations, that analytic solutions are the exception, not the rule, and that solutions sets may show "deterministic chaos."

Linear equations enjoy by definition the property of superposition. That is, linear combinations of solutions are also solutions: the solutions form a linear, or vector, space. Linear theories are highly structured theories, and one has many helpful theorems at hand. For example, a general solution exists; solutions have only "fixed" singularities, that is, those occurring in the linear equations themselves. But do not get the idea that linear theories are considered *passé* or discredited, now that we are elucidating the mysteries of nonlinear dynamics. Some of the most beautiful and accurate theories in physics are linear. Witness the Maxwell theory of electromagnetism, or quantum mechanics itself, the fundamental theory of the subatomic world. Indeed, today no failure of quantum mechanics is known.

Nonlinear equations are all the rest: all those which are *not* linear. Most of the convenient properties of a linear dynamics mentioned above are lost: there is no superposition and no general solution; analytic solutions are rare or nonexistent; solutions may have singularities not present in the motion equations, and these may depend on the initial conditions, etc. However, interesting new properties show up in compensation. Asymptotic (time $\to \infty$) solutions are often independent of initial conditions and lie on low-dimensional "attractors" in phase space. There is a complicated set of these stable regimes joined by bifurcations of various types. There may exist "chaotic" regimes.

Incidentally, the reader should not worry if some of the statements in these preliminary remarks of chapter 1 seem a bit vague or elusive by reason of undefined terms. For now, it is enough that they carry some sort of intuitive meaning. All important terms and concepts will be defined carefully at the proper places in this report.

1.2 The Goals of Nonlinear Dynamics

The dynamics that will be our main focus of attention in this report will be specified by one or several first order ordinary differential equations in time,

$$\dot{x} = f(x,t) \quad , \quad \dot{x} \equiv \frac{dx}{dt} \quad , \tag{1-1}$$

or by a *map* $x \longmapsto F(x)$ which is *iterated*:

$$x, \quad F(x), \quad F[F(x)], \quad F[F[F(x)]], \quad \cdots. \tag{1-2}$$

The set of continuous time solutions, or orbits of (1-1) is called a *flow*, while the set of discrete-time orbits (1-2) is sometimes called a *cascade*. Both f and F depend on parameters which can be varied.

What information do we seek in nonlinear dynamics?

a. The geometry, or more often, topology of the flow (or cascade) as a whole: phase portraits, stable and unstable manifolds, various low-dimensional invariant attracting sets if they exist.

b. Bifurcation points, that is, those parameter values at which the flow "changes qualitatively."

c. The characteristics of "chaotic" flows, and the various paths to chaos which the dynamics admits.

What mathematical tools are available for this search? There exist theorems, far fewer than in linear dynamics, which limit the possibilities in nonlinear dynamics. Numerical computation (sometimes called "experimental mathematics") plays a big role in discovering the information listed above and in suggesting and motivating, if not proving, theorems about a particular dynamics.

1.3 "Chaos"

The quotation marks here signal that a consensus has not yet been reached on the precise definition of this term. This accounts for the many apparent contradictions and fruitless controversies in this subject. Following current custom, we shall mean by "chaos" any or all of the following properties: sensitive dependence on initial conditions, broadband power spectra, decaying correlations, or randomness or unpredictability of orbits as measured by positive algorithmic complexity or entropies of the various kinds. These properties are not all independent.

Of these, *sensitive dependence* (on initial conditions understood), abbreviated in this report as SD, is by far the most important ingredient of "chaos." In fact, in its strong, or exponential, form SD is accepted by most as the definition of chaos. The intuitive meaning of SD is the unpredictability - or uncomputability - in principle of some orbits. That is, inevitable errors in initial conditions, *no matter how small*, may get magnified on computation, so that the computed orbit (or some observable function of the orbit) bears no resemblance to the actual orbit (or function thereof). This has nothing to do with noise or perturbations from outside the system. Sensitive dependence is an intrinsic property of the dynamics in some parameter regimes; it is true "deterministic chaos." Obviously, this bears on the ancient philosophical dichotomy between determinism and chance (and seems at first sight to contradict it!).

In a system which displays "chaos," there may be several sequences of regimes leading to "chaotic" behavior, several "paths to chaos," so to say. The universality of these various paths in systems superficially very different (for example, iterating one-dimensional maps and viscous, incompressible fluid flow) is a surprising theoretical and experimental result.

To give the reader a preliminary feeling for sensitive dependence, this perhaps most important concept of nonlinear dynamics, we shall illustrate it on the simple dynamics of a 1D (one-dimensional) map. The other attributes of "chaos" mentioned above will be covered later in the main text. Consider the particular 1D map $F(x) \equiv \mu x(1-x)$ with $0 \leq x \leq 1$ and $0 < \mu \leq 4$, that is, the iteration scheme $x_{n+1} = \mu x_n(1-x_n)$, $n = 0, 1, 2, 3, \cdots$, defining the orbit (1-2). Choose the parameter value $\mu = 4$ and substitute $x_n \equiv \sin^2 \pi\theta_n$, $0 \leq \theta_n \leq 1$. Then the iteration scheme takes the form $\sin^2 \pi\theta_{n+1} = 4\sin^2 \pi\theta_n \cos^2 \pi\theta_n$, that is,

$$\theta_{n+1} = 2\theta_n(mod1) \qquad (1-3)$$

where $(mod1)$ means that any integral part of $2\theta_n$ is chopped off so that the result lies in the interval (0,1). We can actually get an "analytic" solution (!) for this parameter value, namely

$$\theta_n = 2^n\theta_0(mod1),$$

where $\theta_0 \in (0,1)$ is the initial value. Now shift the initial point slightly: $\theta_0' = \theta_0 + \epsilon$; then

$$\theta_n' - \theta_n = 2^n\epsilon = \epsilon e^{n \ln 2}$$

(as long as $2^n\epsilon < 1$), that is, exponential separation of the two initially very close orbits with *Lyapunov exponent* ln 2. Obviously, the "error" in the orbit will get big for n large enough, no matter how small ϵ. This is SD (in particular, exponential SD).

To be more quantitative about the SD, write the initial value θ_0 in binary notation. For example,

$$\theta_0 = 1/2 + 1/4 + 1/16 + 1/128 + \cdots = 0.1101001\cdots. \qquad (1-4)$$

Then iteration algorithm (1-3) amounts to shifting the "decimal point" to the right by one and dropping the digit to the left of this point. For the value (1-4),

$$\theta_0 = .1101001\cdots, \quad \theta_1 = .101001\cdots, \quad \theta_2 = .01001\cdots, \quad \theta_3 = .1001\cdots, \quad \text{etc.}$$

We see that θ_n depends on the $(n+1)$st and higher digits of θ_0, so when n is large, the value of θ_n depends extremely sensitively on the precise value of θ_0. For instance, let θ_0 and θ'_0 differ first in the $(n+1)$st place, where θ_0 has a 0 and θ'_0 has a 1. Then $\theta'_0 - \theta_0 = 2^{-n}$ at most ($<<< 1$ for large n). But $\theta_n = .0\cdots$ and $\theta'_n = .1\cdots$, so that they could differ by as much as 1, or the whole domain (0,1) of the logistic map for $0 < \mu \leq 4$. On a digital computer with capacity 2^N bits, the computed orbit for a given θ_0 has in general no resemblance to the real orbits for times $n \geq N$.

Ex. 1.1 Take $\theta_0 = 1/7$. Then we know that the exact orbit is

$$1/7, \quad 2/7, \quad 4/7, \quad 1/7, \quad 2/7, \quad 4/7, \quad 1/7\cdots, \qquad (1-5)$$

that is, a periodic orbit of period 3. Now perform the iteration (1-3) on a pocket calculator or computer and compare with (1-5) for large n.

2. FUNDAMENTALS OF CONTINUOUS TIME SYSTEMS

2.1 Flows

A system of N first order *ordinary* differential equations in time t,

$$\dot{x} = f(x,t), \quad \dot{x} \equiv \frac{dx}{dt}, \qquad x \in \Re^N, \tag{2-1}$$

defines a *flow*. Here we have taken the flow to be in $\Re^N \equiv$ the set of all real N-tuples $(x_1, x_2, \cdots x_N)$, which is the usual case; the function f, which thus has N components $(f_1, f_2, \cdots f_N)$, maps $\Re^N$ into $\Re^N$, in symbols $f : \Re^N \to \Re^N$. If $f(x,t) \equiv f(x)$ does not explicitly depend on t, the flow is called *autonomous.* $\Re^N$, or the subset of $\Re^N$ in which the flow is confined, is called the *phase space* of the flow. A solution $x(t) = (x_1(t), x_2(t), \cdots x_N(t))$ of the flow (2-1) with initial value $x_0 \equiv x(0) \equiv (x_1(0), x_2(0), \cdots x_N(0))$ is called the *orbit.* A graph of all orbits or some subset of them in phase space is called a *phase portrait,* and is useful to visualize the flow as a whole in the neighborhood of some interesting point or other structure.

Orbits of an autonomous flow do not intersect! Every point in phase space lies on one and only one orbit. This comes from a beautiful theorem on the uniqueness of orbits, see, say, Guckenheimer and Holmes [20] (Th. 1.0.1), hereafter also GH, which states precisely:

> Let f be C^1 in $\Re^N$. For any open set $U \subset \Re^N$, $\exists$ a time interval $(-c, c)$ such that the orbit $\phi_t(x_0)$ exists and is unique for every $x_0 \in U$. (2 – 2)

For technical mathematical symbols and terms here and hereafter, consult the mathematical Glossary at the end of Guckenheimer and Holmes. We shall use the symbol $\exists$, "there exists," quite often. We shall usually assume the hypotheses of

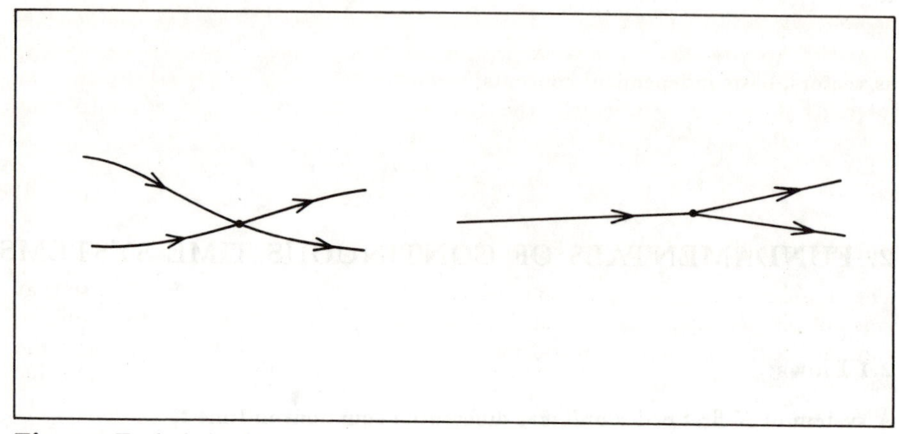

Fig. 2.1 Excluded orbits

this theorem fulfilled for our flows, so that orbits like those shown in Fig. 2.1 are excluded.

The reader might think that restricting our dynamics to autonomous flows (as we shall do) is much too narrow. It seems to rule out higher order motion equations, for example, second order equations such as Newton's laws deliver, all cases with forcing terms, and so on. But this is not so. By enlarging our phase space we can include those cases too. An example will make this clear. Consider the nonautonomous, second order dynamics defined by $\ddot{x}+x = a\cos\omega t$, a harmonically driven linear harmonic oscillator with position coordinate x. Set $x_1 \equiv x$, $x_2 \equiv \dot{x}$, $x_3 \equiv \omega t$. Then we get

$$\dot{x}_1 = x_2, \quad \dot{x}_2 = -x_1 + a\cos x_3, \quad \dot{x}_3 = \omega \quad .$$

But this is just an autonomous flow in $\Re^3$! In particular, even without the forcing term ($a = 0$), the phase space is $\Re^2$, not $\Re^1 \equiv \Re$; phase space is the space of position *and* velocities (or momenta), so it has dimension $2m$ for a configuration space of dimension m. Hence without loss of generality we shall assume all flows autonomous hereafter.

2.2 Linear Stability Analysis

2.2.1 Case of Linear Flows

Consider the *linear* autonomous flow $\dot{x} = Ax$, where A is a real $N \times N$ matrix. We treat the case which *usually* occurs in applications: A can be diagonalized by a similarity transformation

$$T^{-1}AT = \Lambda \quad , \tag{2-3}$$

where Λ is diagonal with the eigenvalues $\lambda_1, \lambda_2, \cdots \lambda_N$ of A on the diagonal. Thus the corresponding eigenvectors $e_1, e_2, \cdots e_N$,

$$Ae_i = \lambda_i e_i \quad , \quad i = 1, 2, \cdots N, \tag{2-4}$$

are *linearly independent* (*span* $\Re^N$), and the columns of T are the components of these eigenvectors. (We prefer to regard A as a linear operator and the e_i as vectors, basis-independent concepts; nevertheless, entirely equivalently one can interpret A as an $N \times N$ matrix, the e_i as $N \times 1$, or column, matrices, and Ae_i as matrix multiplication.) The *secular equation*

$$|A - \lambda \mathbf{1}| \equiv \det(A - \lambda \mathbf{1}) = 0, \qquad (2-5)$$

where $\mathbf{1} \equiv$ unit matrix, determines the eigenvalues λ_i.

The completely general case, when the eigenvectors of A may not span $\Re^N$, so that A is not diagonalizable, is treated in appendix A. A can still be put into a simple, standard form (Jordan canonical form) by a similarity transformation, and the resulting linear stability analysis is not essentially different from the diagonalizable case.

We now define some important subspaces of phase space. Divide the eigenvectors into three subsets,

$$\begin{aligned} &\{u_1, u_2, \cdots u_{N_u}\} \quad \text{such that} \quad Re\lambda_i > 0 \quad , \\ &\{v_1, v_2, \cdots v_{N_s}\} \quad \text{such that} \quad Re\lambda_i < 0 \quad , \\ &\{w_1, w_2, \cdots w_{N_c}\} \quad \text{such that} \quad Re\lambda_i = 0 \quad , \end{aligned} \qquad (2-6)$$

with $N_u + N_s + N_c = N$. Then define

$$\begin{aligned} \text{Unstable subspace} \quad & E^u \equiv span\{u_1, u_2, \cdots u_{N_u}\}, \\ \text{Stable subspace} \quad & E^s \equiv span\{v_1, v_2, \cdots v_{N_s}\}, \\ \text{Center subspace} \quad & E^c \equiv span\{w_1, w_2, \cdots w_{N_c}\}. \end{aligned} \qquad (2-7)$$

The reason for the nomenclature is this: we assert that every orbit based at $x_0 \in E^s$ decays exponentially in t; every orbit based at $x_0 \in E^u$ blows up exponentially in t; and every orbit based at $x_0 \in E^c$ is constant in t, as $t \to +\infty$. We also claim that each subspace is *invariant* (carried into itself) under the flow. Both of these assertions are easily seen by noting that the solution of $\dot{x} = Ax$ is $x(t) = \exp(tA)x_0$. Taking $x_0 \equiv \sum_1^{N_s} c_j \cdot e_j$ in the stable subspace E^s, for example, we see that the orbit is

$$\exp(tA)x_0 = \exp(tA)\sum c_j e_j = \sum c_j \exp(tA) e_j = \sum c_j e^{t\lambda_j} e_j \in E^s \quad , \qquad (2-8)$$

Q.E.D. Moreover, since $Re\lambda_j < 0$, the *length* $\|x(t)\| \longrightarrow 0$. Similarly for $x_0 \in E^u$, $\|x(t)\| \longrightarrow +\infty$; for $x_0 \in E^c$, $\|x(t)\| = \|x_0\| =$ const.

A word on the general case: the three subspaces are defined by (2-6) and (2-7), where the vectors are now generalized eigenvectors to the eigenvalues determined by (2-5). One can show that these subspaces are invariant and that every orbit based in E^s, E^u, or E^c decays exponentially, blows up exponentially, or varies algebraically in t as $t \to +\infty$. The only difference is that powers of t times an exponential in t are in general allowed.

Note that the point $x^* = \mathbf{0}$ (the zero *vector*) is a *fixed point*: $\dot{x}^* = 0$ for the linear flow, in fact the only fixed point. The phase portrait of the flow near the fixed point can be constructed, and the subspaces E^u, E^s, and E^c indicated on the same graph.

Ex. 2.1 Take $A = \begin{pmatrix} 0 & 1 \\ 0 & -4 \end{pmatrix}$. The phase space is $\Re^2 \equiv$ the plane.

a) Find the eigenvalues and eigenvectors. Is A diagonalizable?

b) Find E^u, E^s, and E^c.

c) Draw the phase portrait around $x^* = \mathbf{0} \equiv (0,0)$, and indicate the three subspaces.

Ex. 2.2 Same question for

$$A = \begin{pmatrix} -1 & -1 & 0 \\ 1 & -1 & 0 \\ 0 & 0 & 2 \end{pmatrix}, \quad \text{phase space is } \Re^3.$$

As to part c. of Ex. 2.2, here there is a pair of complex conjugate eigenvalues and eigenvectors: λ_+, e_+ and $\lambda_- = \lambda_+^*, e_- = e_+^*$, where * is complex conjugate. Form two real vectors e_1, e_2 from complex combinations of e_+ and e_-; then e_1, e_2 span the same real two-dimensional subspace of $\Re^3$ as e_+ and e_-. Express the real orbits in terms of e_1 and e_2. You will find spiralling motion.

2.2.2 Case of Nonlinear Flows

The linear stability analysis of the flow $\dot{x} = f(x)$, in general nonlinear, now follows easily from that of linear flows. Consider a *fixed point* x^* of the flow, defined by $\dot{x}^* = f(x^*) = 0$. We *linearize* the flow about x^*. Set $x = x^* + u$, where $\|u\|$ is small in some sense, and keep only terms of $O(u)$ in the calculation. Substitute $x = x^* + u$ into the flow equations, expand $f(x^* + u)$ in a power series in u about x^*, and keep only the first two terms. For the $O(u)$ part we get

$$\dot{u} = Df(x^*)u \quad , \tag{2-9}$$

where $Df(x^*)$ is the *Jacobian matrix* evaluated at the fixed point,

$$[Df(x^*)]_{ij} \equiv \left.\frac{\partial f_i}{\partial x_j}\right|_{x=x^*}, \quad i,j = 1,2,\cdots N. \tag{2-10}$$

But (2-9) is just a linear flow with $A = Df(x^*)$. So we find the eigenvalues and eigenvectors, invariant subspaces E^u, E^s, E^c, etc.; that is, we perform the *linear* *stability analysis* for this fixed point just as in section 2.2.1. We expect the local nonlinear flow around x^* to be indistinguishable from the linear flow governed by $A \equiv Df(x^*)$. This is true with an important proviso to be made below.

We work out an illustrative example: consider the *van der Pol oscillator* $\ddot{x} + b(x^2 - 1)\dot{x} + x = 0$, $b > 0$. As an autonomous flow in $\Re^2$ it reads $\dot{x}_1 = x_2$, $\dot{x}_2 = -x_1 - b(x_1^2 - 1)x_2$, so $f_1 = x_2$, $f_2 = -x_1 - b(x_1^2 - 1)x_2$. The only fixed point is $x^* = (0,0)$. The partial derivatives of f are

$$\frac{\partial f_1}{\partial x_1} = 0, \quad \frac{\partial f_1}{\partial x_2} = 1, \quad \frac{\partial f_2}{\partial x_1} = -1 - 2bx_1x_2, \quad \frac{\partial f_2}{\partial x_2} = -b(x_1^2 - 1),$$

so the Jacobian matrix is

$$Df(x^* = 0) = \begin{pmatrix} 0 & 1 \\ -1 & b \end{pmatrix}. \tag{2-11}$$

The eigenvalues of this are $\lambda_\pm = b/2 \pm (b^2/4 - 1)^{\frac{1}{2}}$, and the corresponding eigenvectors $e_\pm$ are found to span $\Re^2$. Now $Re\lambda_\pm > 0$, so

$$E^u = span\{e_+, e_-\} = \Re^2, \quad E^s = E^c = \mathbf{0}. \tag{2-12}$$

All orbits are repelled exponentially from the fixed point $\mathbf{0}$.

2.3 Stability Types of Fixed Points

We now have to elucidate the key notion of stability (for a fixed point here, but for more general structures later), and to see what linear stability analysis has to say about it. For the fixed point x^* of a general autonomous flow, which we assume is confined to the open set $U \subset \Re^N$, we have the definitions (GH, p.3):

> The fixed point x^* is *stable* if for every neighborhood $V \subset U$ of x^* there is a neighborhood $V_1 \subset V$ of x^* such that every solution $x(t) = \phi_t(x_0)$ with $x_0 \in V_1$ is defined and $\in V$ for all $t > 0$. (2 – 13a)

> The fixed point x^* is *asymptotically stable* if it is stable and for every neighborhood $V \subset U$ of x^* a neighborhood $V_1 \subset V$ of x^* exists such that $\phi_t(x_0) \to x^*$, $t \to +\infty$, for every $x_0 \in V_1$. (2 – 13b)

> The fixed point x^* is *unstable* if it is not stable. (2 – 13c)

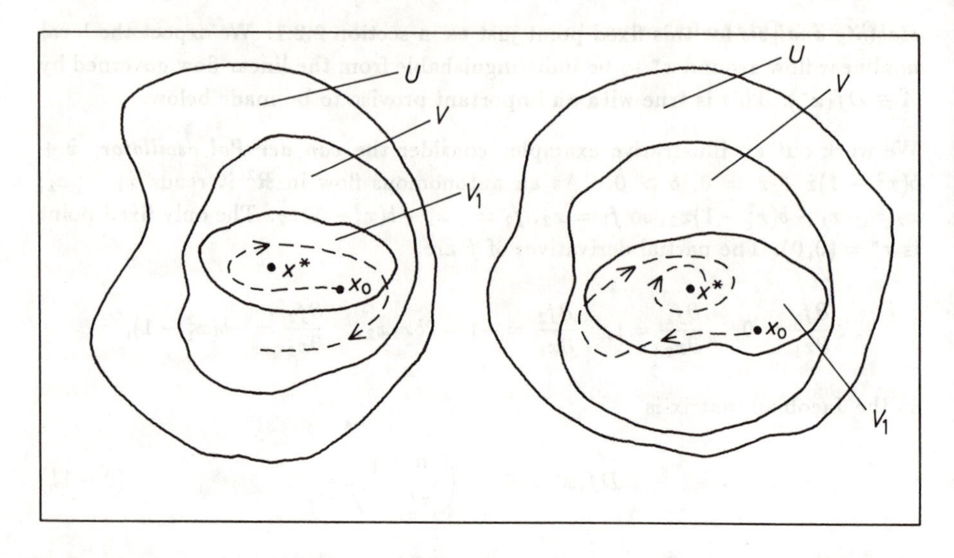

Fig. 2.2 Stability types illustrated: on the left, a center (stable but not asymptotically stable); on the right, a sink (asymptotically stable)

See Fig. 2.2.

The fixed point has special names in the first two cases: x^* is a *center* if it is stable but not asymptotically stable; it is a *sink* if it is asymptotically stable. It is useful to spell out the definition of unstable in positive terms by negating the definition of stable. Namely:

> The fixed point x^* is *unstable* if $\exists$ a neighborhood $V \subset U$ of x^* such that for all neighborhoods $V_1 \subset V$ of x^* $\exists$ an orbit based at $x_0 \in V_1$ which $\notin V$ for some $t > 0$. (2 – 14)

Loosely transcribed, these careful definitions can be phrased as follows. A fixed point x^* is stable if for every neighborhood V of x^*, we can keep all orbits in V forever if we start them close enough to x^*. It is asymptotically stable if it is stable and if all orbits converge to x^* if we start them close enough to x^*. It is unstable if we can find a neighborhood V of it such that from any neighborhood V_1 of x^* contained in V, no matter how small, at least one orbit started in V_1 escapes from V at some time.

To convince yourself of the necessity of the details which seem to make the careful definitions (2-13) unnecessarily pedantic and complicated, try answering the questions in Ex. 2.3.

Ex. 2.3 Do the following statements say anything about the stability type of x^*?

a) For every neighborhood V of x^* an orbit based in V converging to x^* is observed.

b) For every neighborhood V of x^* an orbit based in V which escapes V is observed.

c) We can find a neighborhood V of x^* such that for every neighborhood $V_1 \subset V$ of x^*, every orbit based in V_1 stays in V for all $t > 0$.

2.4 Connection of Stability and Linear Stability Analysis

The general idea is that the local behavior of the linearized flow around x^* carries over to the nonlinear flow, thus selecting one of its stability types. To make this precise, let us call x^* *hyperbolic* (sometimes *nondegenerate*) if $Df(x^*)$ has no eigenvalue λ with $Re\lambda = 0$. That is, if $E^c = \mathbf{0}$. If then $E^s = \mathbf{0}$, x^* is called a *source*; if both E^s and $E^u \neq \mathbf{0}$, x^* is called a *saddle point.* See Fig. 2.3.

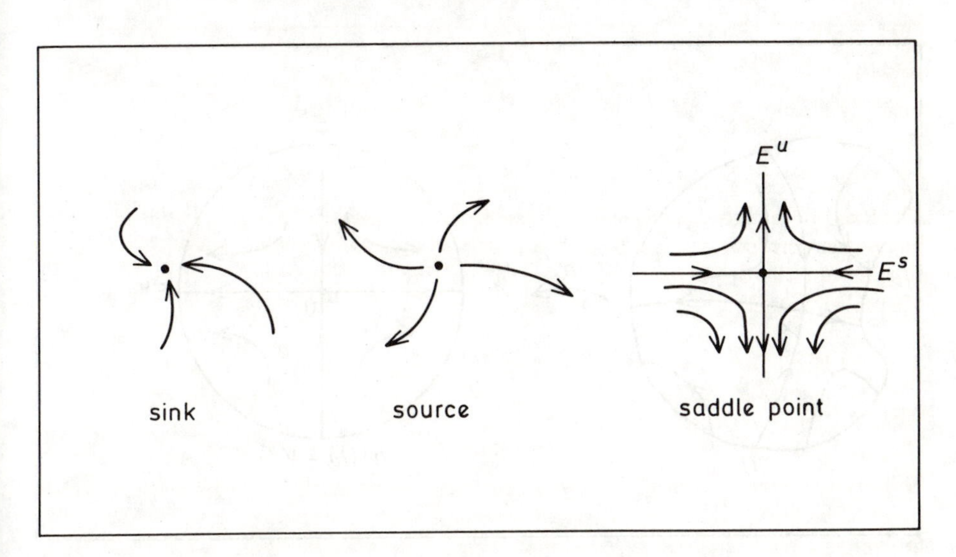

Fig. 2.3. Left to right: a sink, a source, a saddle point

Then there exists a *homeomorphism h* in some neighborhood V of x^* taking the orbits $\phi_t(x_0)$ of the nonlinear flow into the orbits $\exp(tDf(x^*))u_0$ of the linearized flow and preserving the sense of the orbits (GH, Th 1.3.1).

Def. A *homeomorphism* $h : U \to V$ maps open set U into open set V such that h is continuous and h^{-1} exists and is continuous. $\qquad (2-15)$

See Fig. 2.4.

Homeomorphism is the basic concept of topology. A topological property is one invariant under all homeomorphisms (thus under stretching, compressing, twisting, etc., but not tearing). The importance of homeomorphisms and topology in nonlinear dynamics is that *stability is a topological notion!*

Thus in the hyperbolic case the linearization determines the asymptotic time behavior of the nonlinear flow *and hence x^*'s stability type.* We have

x^* is asymptotically stable iff only $E^s \neq 0$ (x^* a *sink*).

x^* is unstable iff $E^u \neq 0$ (x^* is a *source* or a *saddle point*).

(Hyperbolic case) (2 − 16)

Iff ≡ "if and only if." The *center* case is absent. These are the only two possibilities. If $Re\lambda = 0$ for any eigenvalue of the Jacobian matrix ($E^c \neq 0$), the stability of x^* *cannot* in general be determined by the linearization.

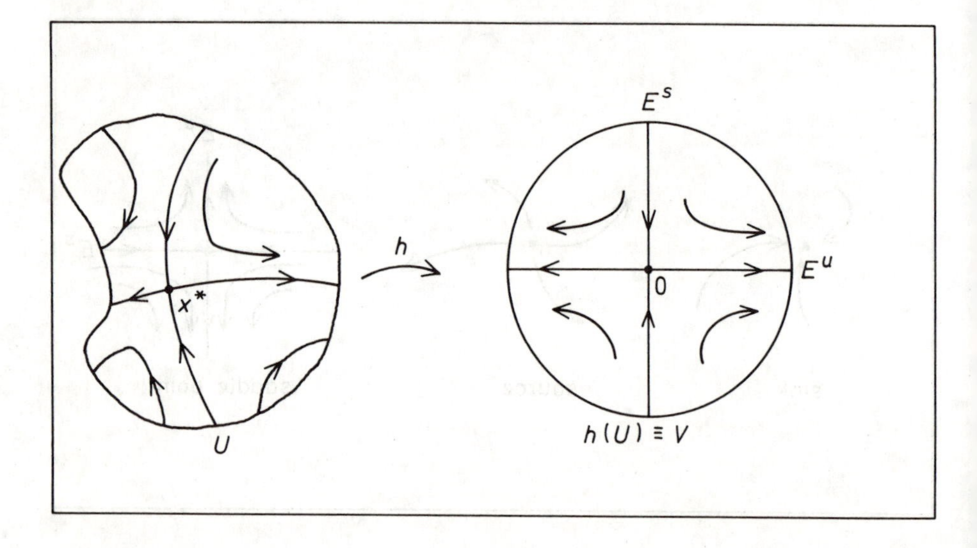

Fig. 2.4 Homeomorphism from the nonlinear to the linearized flow around a fixed point.

2.5 Topological Equivalence of Flows

The above section motivates the introduction of another topological notion which will be useful when we come to discuss bifurcations. First we give the rigorous definition, then try to give the reader an intuitive feeling for it by some remarks.

Consider two flows in $\Re^N$, $\dot{x} = f(x)$, $x \in X$, and $\dot{y} = g(y), y \in Y$, with orbit functions ϕ_t^f and ϕ_t^g, respectively.

Def. The f-flow and the g-flow are *topologically equivalent* iff there exists a homeomorphism $h : X \to Y$ such that for every t_1,
$$h \circ \phi_{t_1}^f = \phi_{t_2}^g \circ h \text{ for some } t_2. \tag{2-17}$$

Here $\circ$ is functional composition, namely $f \circ g(x) \equiv f(g(x))$. If the homeomorphism is realized by the function $y = h(x)$, this means

$$h(\phi_{t_1}^f(x)) = \phi_{t_2}^g(h(x)), \quad \text{all} \quad x \in X. \tag{2-18}$$

That is, there exists a 1-1 bicontinuous map which takes every point x on an orbit of the f-flow at time t_1 into a point $y = h(x)$ of the orbit of the g-flow at a time t_2. See Fig. 2.5.

More briefly, h maps every orbit of the f-flow into an orbit of the g-flow in a continuous manner, and vice-versa. Stated yet another way, the f-flow can be *continuously deformed* into the g-flow, and vice-versa. Thus the two flows are really the same up to a "change of coordinates," as (2-17) validates. In this language the theorem in section 2.4 says that the nonlinear flow is topologically equivalent to the linearized flow around a hyperbolic fixed point.

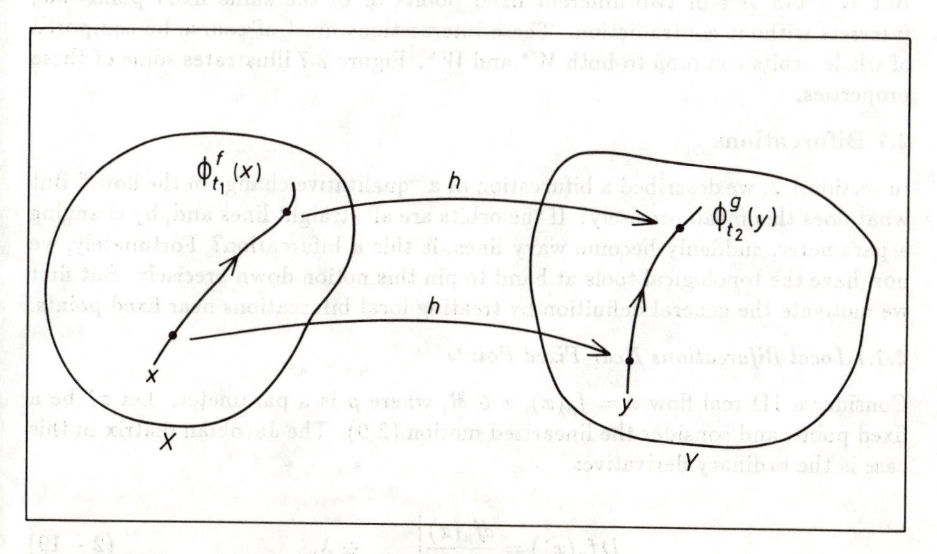

Fig. 2.5 Mapping of topologically equivalent flows

2.6 Stable and Unstable Manifolds of a Fixed Point

These are generalizations of the stable and unstable subspaces of the linearized flow to the full nonlinear flow. We shall give loosely stated definitions sufficient for our purposes.

The (global) stable manifold W^s of a hyperbolic fixed point x^* is the set of all those points which converge to x^* under the flow. That is, all $x \in$ phase space such that $\phi_t(x) \to x^*, \quad t \to +\infty$. Similarly, the (global) unstable manifold W^u of x^* is the set of all those points which "diverge from x^* under the flow" or, more precisely, all those points which converge to x^* as time runs backwards: $\phi_t(x) \to x^*, \quad t \to -\infty$. Both W^s and W^u are invariant sets, where the general definition is: set S is *invariant* if $\phi_t(S) \subset S, \quad -\infty < t < \infty$. See Fig. 2.6, which illustrates W^s and W^u, both 1D (one-dimensional) for a 2D flow, and also their relations to E^s and E^u.

Then we have the Stable Manifold Theorem (GH, Th 1.3.2), which states that W^s and W^u have the same dimensions as E^s and E^u, respectively, and are tangent to them at the hyperbolic fixed point x^*. Figure 2.6 illustrates this.

We detail some properties of W^s and W^uin the following. Since an orbit is an invariant set and orbits do not intersect, we see that W^s and W^u are composed of entire orbits. Clearly, given two (hyperbolic understood) fixed points x^*, y^*, $W^s(x^*)$ and $W^s(y^*)$ cannot intersect, by definition. The same goes for $W^u(x^*)$ and $W^u(y^*)$.

But W^s and W^u of two different fixed points or of the same fixed point *may intersect* without contradiction. These intersections must of course be composed of whole orbits common to both W^s and W^u. Figure 2.7 illustrates some of these properties.

2.7 Bifurcations

In section 1.2, we described a bifurcation as a "qualitative change in the flow." But what does this mean precisely? If the orbits are all straight lines and, by changing a parameter, suddenly become wavy lines, is this a bifurcation? Fortunately, we now have the topological tools at hand to pin this notion down precisely. But first we motivate the general definition by treating local bifurcations near fixed points.

2.7.1 Local Bifurcations Near Fixed Points

Consider a 1D real flow $\dot{x} = f_\mu(x)$, $x \in \Re$, where μ is a parameter. Let x^* be a fixed point, and consider the linearized motion (2-9). The Jacobian matrix in this case is the ordinary derivative:

$$Df_\mu(x^*) = \left. \frac{df_\mu(x)}{dx} \right|_{x=x^*} \equiv \lambda, \qquad (2-19)$$

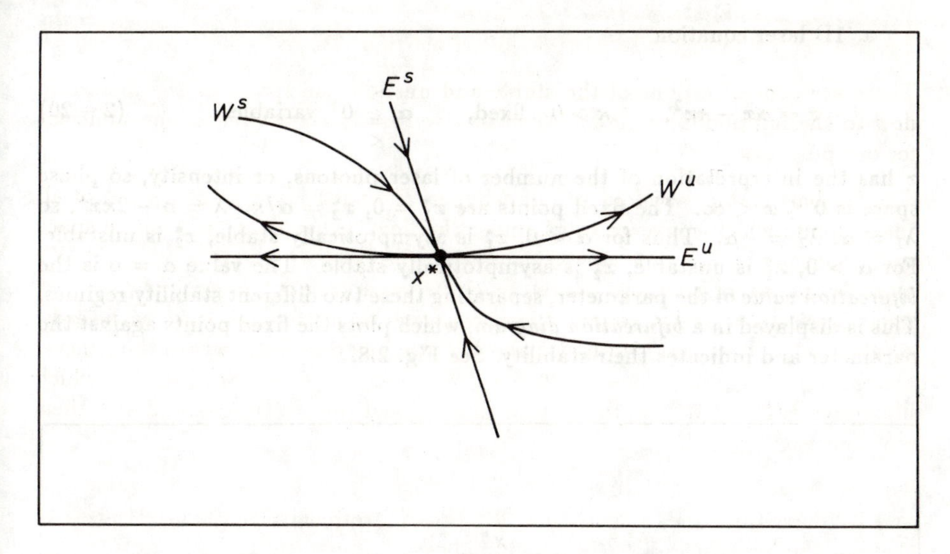

Fig. 2.6 Relation of W^s and E^s and of W^u, E^u of a fixed point.

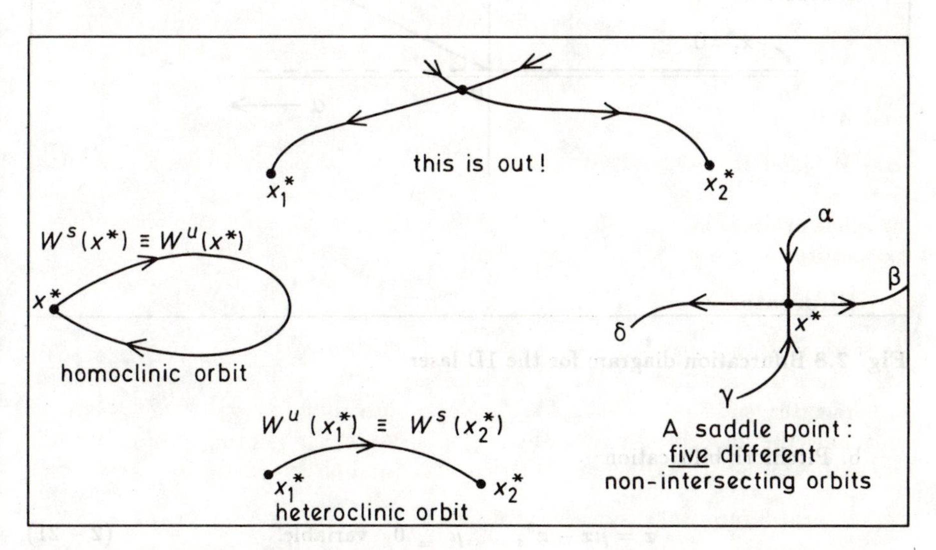

Fig. 2.7 Top: forbidden transverse intersection of stable manifolds. Bottom: allowed cases of stable and unstable manifolds.

the eigenvalue itself. Thus in the case $\lambda \neq 0$ (note λ is real) we have the hyperbolic case, and the stability type of x^* is completely determined by the principle (2-16). Namely, x^* is asymptotically stable if $\lambda < 0$ and unstable if $\lambda > 0$.

a. 1D laser equation

$$\dot{x} = \alpha x - \kappa x^2, \qquad \kappa > 0 \quad \text{fixed}, \qquad \alpha \underset{<}{\overset{>}{=}} 0 \quad \text{variable.} \tag{2-20}$$

x has the interpretation of the number of laser photons, or intensity, so phase space is $0 \leq x < \infty$. The fixed points are $x_1^* = 0$, $x_2^* = \alpha/\kappa$. $\lambda = \alpha - 2\kappa x^*$, so $\lambda_1 = \alpha$, $\lambda_2 = -\alpha$. Thus for $\alpha < 0$, x_1^* is asymptotically stable, x_2^* is unstable. For $\alpha > 0$, x_1^* is unstable, x_2^* is asymptotically stable. The value $\alpha = 0$ is the *bifurcation value* of the parameter, separating these two different stability regimes. This is displayed in a *bifurcation diagram*, which plots the fixed points against the parameter and indicates their stability. See Fig. 2.8.

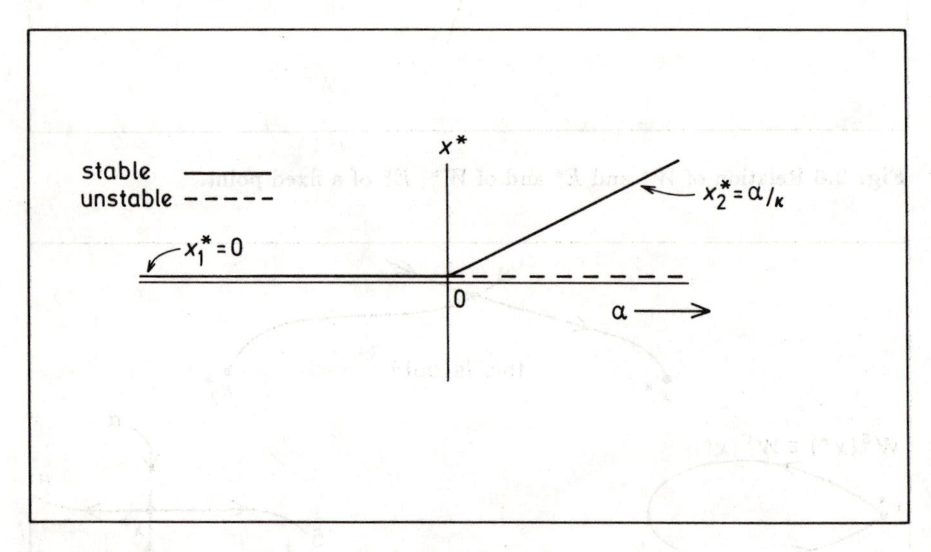

Fig. 2.8 Bifurcation diagram for the 1D laser

b. Pitchfork bifurcation

$$\dot{x} = \mu x - x^3, \qquad \mu \underset{<}{\overset{>}{=}} 0 \quad \text{variable.} \tag{2-21}$$

Here $-\infty < x < \infty$. The fixed points are $x_1^* = 0$, $x_\pm^* = \pm\sqrt{\mu}$. $\lambda = \mu - 3(x^*)^2$, so $\lambda_1 = \mu$, $\lambda_\pm = -2\mu$. Hence for $\mu < 0$, x_1^* is asymptotically stable and the fixed points $x_\pm^*$ don't exist (since phase space is real). For $\mu > 0$, x_1^* is unstable, $x_\pm^*$ are asymptotically stable. We see that $\mu = 0$ is the bifurcation value, and the bifurcation diagram has the form in Fig. 2.9.

The linearized motion in the 1D case is $u = e^{\lambda t}u_0$, so the phase portraits are as shown in Fig. 2.10 in the different stability regimes.

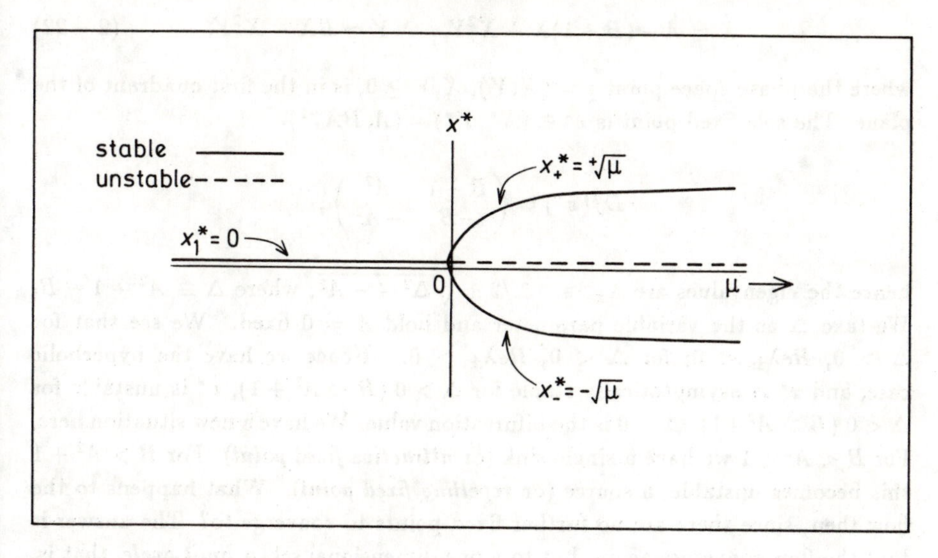

Fig. 2.9 Pitchfork bifurcation diagram

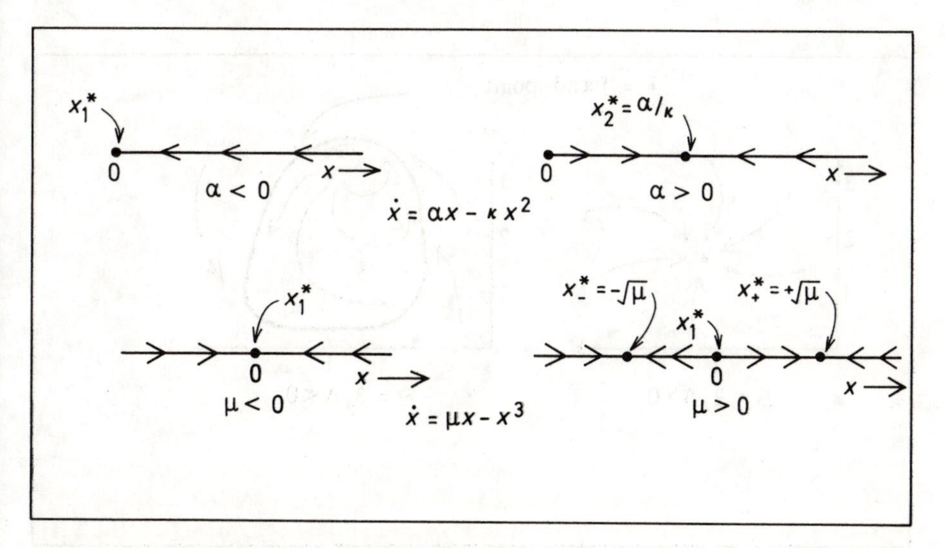

Fig. 2.10 Phase portraits for the 1D flow (2-21)

c. The Brusselator (a model chemical reaction)

$$\dot{X} = A - (B+1)X + X^2Y, \qquad \dot{Y} = BX - X^2Y, \qquad (2-22)$$

where the phase space point $x = (X, Y)$, $X, Y \geq 0$, is in the first quadrant of the plane. The sole fixed point is $x^* \equiv (X^*, Y^*) = (A, BA^{-1})$.

$$Df(x^*) = \begin{pmatrix} B-1 & A^2 \\ -B & -A^2 \end{pmatrix},$$

hence the eigenvalues are $\lambda_\pm \equiv -\Delta/2 \pm \sqrt{\Delta^2/4 - A^2}$, where $\Delta \equiv A^2 + 1 - B$. We take Δ as the variable parameter and hold $A \neq 0$ fixed. We see that for $\Delta > 0$, $Re\lambda_\pm < 0$; for $\Delta < 0$, $Re\lambda_\pm > 0$. Hence we have the hyperbolic case; and x^* is asymptotically stable for $\Delta > 0$ $(B < A^2 + 1)$, x^* is unstable for $\Delta < 0$ $(B > A^2+1)$. $\Delta = 0$ is the bifurcation value. We have a new situation here. For $B < A^2 + 1$ we have a single sink (or *attractive fixed point*). For $B > A^2 + 1$ this becomes unstable, a source (or *repelling fixed point*). What happens to the flow then, since there are no further fixed points to converge to? The answer is that the flow converges again, but to a one-dimensional set, a *limit cycle*, that is, a periodic orbit. See the phase portraits in Fig. 2.11.

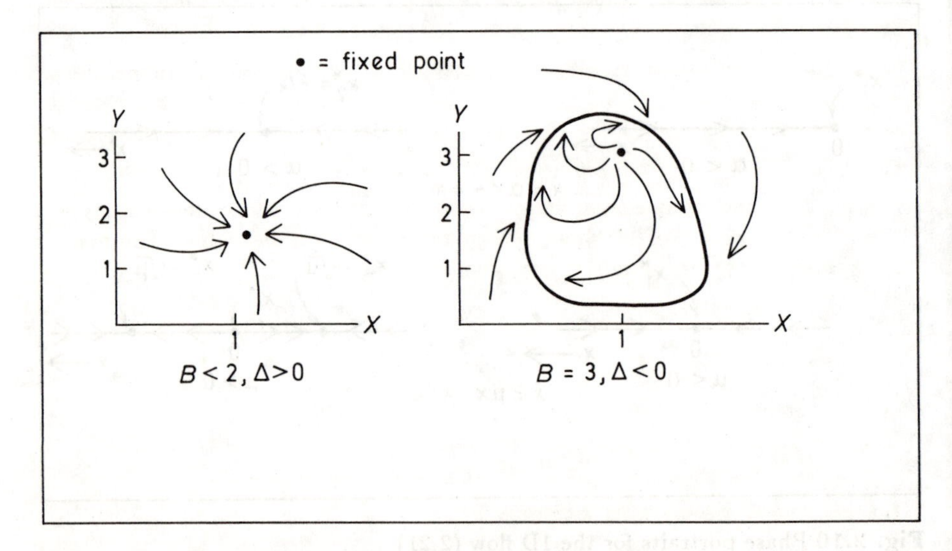

Fig. 2.11 Phase portraits for the Brusselator (A=1)

Motivated by these examples, we generalize to the definition: a *local bifurcation* at the fixed point x^* of the flow $f_\mu(x)$ occurs when an eigenvalue λ of $Df_\mu(x^*)$ crosses the imaginary axis in the complex eigenvalue plane. Let it cross at the *bifurcation value* μ_b of the parameter. Then $Re\lambda(\mu_b) = 0$, and this zero is isolated.

2.7.2 Bifurcation, General Definition

Look at the phase portraits in Figs. 2.10 and 2.11. The flows are really "qualitatively different" before and after the bifurcation. What we mean is that there is no way we can continuously deform one flow into the other. This is the key idea, bringing in the notion (2-17) of topological equivalence.

Def. The parameter value μ_b of the flow $\dot{x} = f_\mu(x)$ is called a *bifurcation value* if the flow is *not structurally stable* at $\mu = \mu_b$. This means that there exist arbitrarily small perturbations $\delta_1 f(x)$, $\delta_2 f(x)$ of $f_{\mu_b}(x)$ such that the perturbed flow $\dot{x} = f_{\mu_b}(x) + \delta_1 f(x)$ is not topologically equivalent to the perturbed flow $\dot{x} = f_{\mu_b}(x) + \delta_2 f(x)$. (2 − 23)

This is still not completely unambiguous, since the nature of these perturbations must be spelled out (see GH, definitions 1.7.1 and 1.7.4), but will suffice for us. Clearly this is what happened in the local bifurcations of section 2.7.1. The small perturbations $\delta_1 f$ and $\delta_2 f$ were generated by changing the parameter μ itself to values slightly above and slightly below the bifurcation value μ_b. Later we shall meet *global* bifurcations, whose modification of the flow is not confined to the neighborhood of any point in phase space, that is, "extends to infinity."

The notion of structural stability of a flow is a very useful one in nonlinear dynamics. For example, if we know that a flow *is* structurally stable at a value μ, we can change μ slightly and be sure that the phase portrait is roughly the same, has not suffered any catastrophic, qualitative changes.

Ex. 2.4 Find fixed points, their stability, and bifurcation values (if any) for the following real 1D flows. Draw the bifurcation diagrams. The real parameter μ can be positive, negative, or zero.

a) $f_\mu(x) \equiv \mu - x^2$

b) $f_\mu(x) \equiv \mu x - x^2$

c) $f_\mu(x) \equiv \mu^2 x - x^3$

d) $f_\mu(x) \equiv \mu^2 x + x^3$

2.8 Dissipative Flows and Attractors

Consider an autonomous flow $\dot{x} = f(x)$ in $\Re^N$. The system is called *dissipative* in $U \subset \Re^N$ if $\nabla \cdot f \equiv \sum_{i=1}^N \frac{\partial f_i}{\partial x_i} < 0$ there. For $f(x)$ is the velocity field of the flow in phase space, so that negative divergence implies that comoving volume elements are shrinking in time. Hence a nonzero volume of $\Re^N$ asymptotically shrinks to volume 0 under the flow. This fact, that flows which initially occupy high-dimensional manifolds eventually end up on very low-dimensional manifolds for dissipative flows, is one of the characteristic and simplifying properties of these nonlinear flows. Such final sets are called attracting sets or attractors. We can make this precise by

Def. A closed and invariant set Λ is called an *attracting set* if $\exists$ a neighborhood V of Λ such that $\phi_t(x) \in V$ for $t \geq 0$ and $\phi_t(x) \to \Lambda$, $t \to +\infty$, for all $x \in V$. (2 – 24)

One can strengthen the definition by requiring some extra properties such as indecomposability (Λ contains a dense orbit), generalized dimension in some range, "chaotic" flow, etc., and call such sets *attractors* or *strange attractors.* We shall not try to be too precise here because there is not yet universal agreement on these definitions in the literature.

Examples of attractors are attractive fixed points and limit cycles. Remember the definition of the former in (2-13b). The notion of the convergence of an orbit to a *set* Λ such as a limit cycle, which occurs in the definition (2-24), is no more complicated than the usual "ϵ, δ" definition of convergence to a point; one must only use the distance of the orbit point $\phi_t(x)$ from the (closed) set Λ rather than the more familiar distance from a point x^*.

Such attractors have "volumes" (that is, Lebesgue measure) in $\Re^N = 0$, and thus ordinary dimension an integer $\leq N - 1$, when they are not too pathological to be assigned an ordinary dimension. For example, an attracting fixed point has dimension 0; an attracting limit cycle, dimension 1. But generalized, *nonintegral* dimension, such as *Hausdorff dimension* HD, may be assigned to any point set in $\Re^N$. Sets with zero volume may have HD any real number between 0 and N. For details, see for example, Young [37]. In fact, requiring $N - 1 < HD(\Lambda) < N$ is a popular criterion that the attractor Λ be strange.

As we saw in section 2.1, orbits do not intersect. This must naturally have some consequences in the cramped phase spaces of one or two dimensions. Such theorems are called generally "no-pass" theorems. They are too technical to quote in this report; the reader is referred to GH, section 1.8, or to Wiggins [36] (section I.1.1) for details. But the general moral of these theorems is that 1D and 2D flows cannot be "chaotic." It is generally agreed that phase space dimension $N \geq 3$ is *necessary* for "chaos." We remind the reader that "chaos" is to be understood as explained in chapter 1; a precise definition will be suggested later.

3. DISCRETE TIME SYSTEMS: ITERATION OF MAPS

3.1 Iteration of Maps

Consider a function, or *map*, F mapping $\Re^N$ into $\Re^N$ which is continuous with continuous first derivatives. We write $F : \Re^N \to \Re^N$, $F \in C^1$. It determines a dynamics in which "time" is discrete by *iteration*, namely

$$x_{n+1} = F(x_n), \qquad n = \text{ integer.} \tag{3-1}$$

We shall occasionally call the map dynamics (3-1) a *cascade*. Thus the nth point of the orbit in terms of the initial point x_0 is

$$x_n = F^n(x_0), \qquad \text{where} \quad F^n(x_0) \equiv \overbrace{F \circ F \circ F \circ \cdots \circ F(x_0)}^{n \text{ times}}$$
$$\equiv F(F(F(\cdots F(x_0)\cdots))), \quad F^0(x_0) \equiv x_0. \tag{3-2}$$

The symbol $F \circ G$ denotes functional composition $F \circ G(x) \equiv F(G(x))$ (don't confuse F^n, the nth iterate, with the nth power).

Note that $F^{n\prime}(x_0) = F'(x_{n-1})F^{n-1\prime}(x_0)$, where prime means derivative, by the chain rule of differentiation, if the phase space is $\Re$. By induction,

$$F^{n\prime}(x_0) = F'(x_{n-1})F'(x_{n-2})\ldots F'(x_0) \quad \text{(Chain Rule).} \tag{3-3}$$

In the case of $\Re^N$, replace $F' \equiv dF/dx$ and $F^{n\prime}$ by DF and DF^n, the Jacobian matrices (2-10), and (3-3) implies matrix multiplication.

Again, orbits do not intersect under certain lenient conditions, cf. the theorem (2-2) for flows. If F has an inverse F^{-1} which is C^1, the orbit based at x_0 is the set (3-1) for $-\infty < n < \infty$; if F is, however, noninvertible, (3-1) with $n = 0, 1, 2, \cdots$ is the orbit. In (3-2) and (3-3) n was tacitly taken positive.

3.2 Linear Stability Analysis

3.2.1 Linear Maps

If $F(x) = Ax$, $A \equiv N \times N$ real matrix, the map is linear. If we can diagonalize A, we define the unstable, stable, and center subspaces by dividing the eigenvectors into the three subsets

$$\begin{aligned} &\{u_1, u_2, \cdots u_{N_u}\} \quad \text{such that} \quad |\lambda_i| > 1, \\ &\{v_1, v_2, \cdots v_{N_s}\} \quad \text{such that} \quad |\lambda_i| < 1, \qquad (3-4) \\ &\{w_1, w_2, \cdots w_{N_c}\} \quad \text{such that} \quad |\lambda_i| = 1, \end{aligned}$$

with $N^u + N^s + N^c = N$. E^u, E^s, E^c are invariant, and vectors based in them blow up, decay, or vary "algebraically" in length respectively as $n \to +\infty$. For now the explicit orbit solution is $x_n = A^n x_0$. Hence for $x_0 = \sum c_j e_j$ belonging to one of these subspaces,

$$x_n = A^n \sum c_j e_j = \sum c_j \lambda_j^n e_j, \qquad (3-5)$$

from which both assertions follow.

Again, the general case A not diagonalizable is not essentially different. See appendix A.

> **Exs. 3.1** and **3.2**. Same as Exs. 2.1 and 2.2, but let the given matrices A define linear *map* dynamics.

3.2.2 Nonlinear Maps

Again, we can use the linear stability analysis for linear maps by linearizing about a *fixed point* x^* of the map $F : F(x^*) = x^*$. Set $x_0 = x^* + u$, $\|u\|$ small, and $x_n \equiv x^* + u_n$. Substituting into (3-2), expanding, and keeping only terms up to O(u), we get

$$\begin{aligned} x_n &= F^n(x^* + u) \approx F^n(x^*) + DF^n(x^*)u = x^* + [DF(x^*)]^n u \\ \Rightarrow u_n &= [DF(x^*)]^n u \end{aligned} \qquad (3-6)$$

We used $F^n(x^*) = x^*$ and $DF^n(x^*) = [DF(x^*)]^n$ from the chain rule (3-3), since $x_0 = x_1 = x_2 \cdots = x_{n-1} = x^*$ for a fixed point. But (3-6) is just the orbit for the linear flow with $A \equiv DF(x^*)$. So the linear stability analysis for the fixed point is performed just as in section 3.2.1.

3.3 Stability Types of a Fixed Point

These are defined just as in section 2.3 with the changes $\phi_t(x_0) \to F^n(x_0)$, $t \to n$.

3.4 Connection of Stability and Linear Stability Analysis

We now call a fixed point of a map F *hyperbolic* if $DF(x^*)$ has no eigenvalue λ with $|\lambda| = 1$, or if $E^c = \mathbf{0}$. *Sources* and *saddle points* are defined by the same language as in section 2.4 where now, of course, the invariant subspaces are identified by (3-4). For x^* hyperbolic there exists a homeomorphism mapping the orbits of the nonlinear map onto those of the linearized map in some neighborhood of x^* and preserving their sense. Thus in the hyperbolic case the linearization determines the stability type, and we get the same two cases: x^* is asymptotically stable (a *sink*) iff only $E^s \neq \mathbf{0}$; it is unstable (a source or saddle point) iff $E^u \neq \mathbf{0}$. Again, if E^c is not $\mathbf{0}$, linear stability does *not* in general determine the stability type of x^*.

3.5 Topological Equivalence of Map Dynamics

Given two cascades (map dynamics) defined by $F(x)$, $x \in X$, and $G(y)$, $y \in Y$,

> Def. The F-cascade and G-cascade are *topologically equivalent* iff there exists a homeomorphism $h : X \to Y$ such that the composition $h \circ F = G \circ h$. $(3-7)$

This is simpler than the definition (2-17) for a flow. This guarantees that every orbit of F is mapped into one of G in a continuous manner, and vice-versa. The F-cascade can be continuously deformed into the G-cascade and vice-versa: they are not really different from a topological point of view. In particular, the nonlinear cascade is topologically equivalent to the linearized cascade around a hyperbolic fixed point.

3.6 Stable and Unstable Manifolds of a Fixed Point

These are defined by exactly the same language as those for flows, section 2.6, with the substitutions $\phi_t(x) \to F^n(x)$ and $t \to n$. The Stable Manifold theorem reads the same. N.B.; W^s and W^u for a map are *manifolds* ("continuous" sets like curves, surfaces, volumes, etc., or technically: locally just like $\Re^M$ for $1 \leq M \leq N$) even though the orbits are discrete point sets. This, of course, is because we can start this hopping orbit anywhere in phase space $\subset \Re^N$. W^s and W^u for maps have the same properties as those detailed in section 2.6.1 for flows. But notice an important difference. A nonempty intersection $W^s \cap W^u$ for flows which is not a fixed point must be at least 1D since they must intersect in whole orbits, which

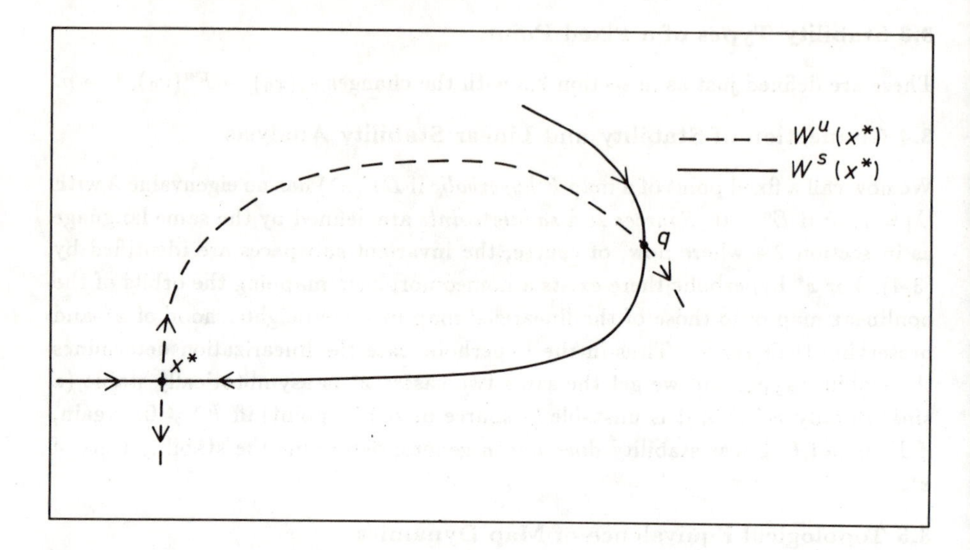

Fig. 3.1 Transverse homoclinic point q.

are 1D manifolds (curves). However, a nonempty $W^s \cap W^u$ for a map can be 0D, a point q which is not a fixed point, as depicted in Fig. 3.1.

Let us study this situation a little more because of its great importance in the subject of "chaos." We consider an orientation preserving map F, in $\Re^2$ for simplicity, with hyperbolic fixed point x^* whose stable and unstable manifold curves W^s and W^u intersect *transversely* in a nonfixed point q (Fig. 3.1). Such a q is called a *transverse homoclinic point* (to x^* understood). Consider the orbit based at q,

$$\{\cdots, F^{-2}(q), F^{-1}(q), q, F(q), F^2(q), \cdots\}. \qquad (3-8)$$

Then since $q \in W^s \cap W^u$ and *these manifolds are both invariant*, the infinite set of points (3-8) must also lie in *both* W^s and W^u. Therefore W^s and W^u must wind between each other (must be "interleaved") in a complicated way, intersecting *at least* in the infinite point set (3-8). These intersections get closer and closer without limit because $F^n(q) \to x^*$ and $F^{-n}(q) \to x^*$ as $n \to +\infty$ by the very definition of W^s and W^u, respectively. Such an orbit (3-8) is called a *transverse homoclinic orbit*, and the complicated geometry of the interleaved, infinitely often intersecting W^s and W^u is named a *homoclinic tangle.* So Fig. 3.1 leads inevitably to Fig. 3.2.

It can be proved that the map dynamics is chaotic in a well-defined sense in a neighborhood of any transverse homoclinic point, (see Wiggins [36], section 4.4, top p. 471). We will return to this notion of chaos in Chap. 5.

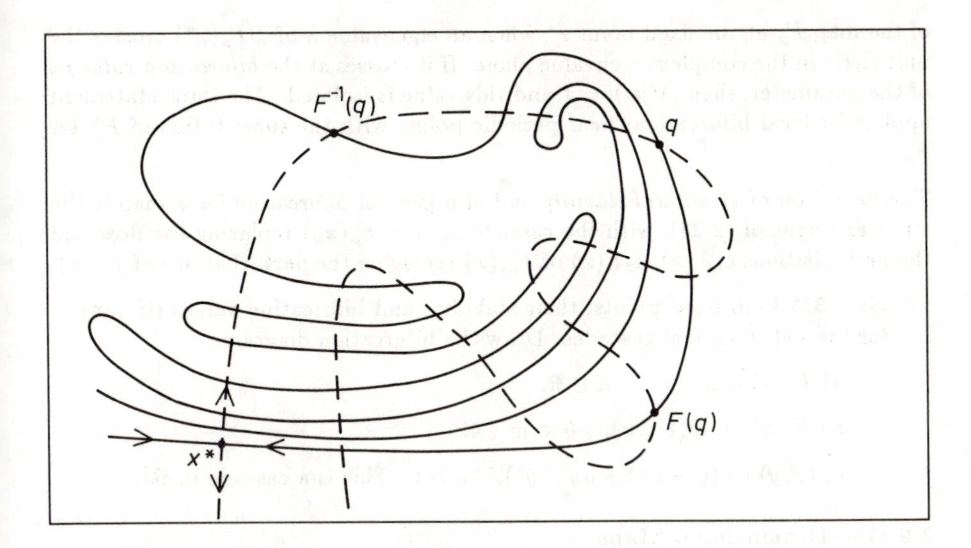

Fig. 3.2 Homoclinic tangle

3.7 Periodic Orbits

If we have a point x_0 such that $F^n(x_0) = x_0$ but $F^m(x_0) \neq x_0$, $m < n$, we see that the n points

$$x_1 = F(x_0), x_2 = F(x_1), \cdots, x_n = F(x_{n-1}) = x_0$$

form a *periodic orbit* of period n (or $n-$ *cycle*, or *fixed point of order n*). Each x_i is a fixed point of $F^n : F^n(x_i) = x_i$, $i = 0, 1, \cdots, n-1$. The stability of the n-cycle can thus be discussed in terms of the stability of the fixed points of the map F^n, a subject which we have already covered, and is fully equivalent to it. Note that

$$DF^n(x_i) = DF(x_0)DF(x_1)\cdots DF(x_{n-1}), \quad i = 0, 1, \cdots, n-1,$$

by the chain rule (3-3) for $\Re^N$, so every fixed point of F^n belonging to an n-cycle has the same Jacobian matrix. So we end up with the characterization of the stability of an n-cycle in the *hyperbolic case* (no eigenvalue λ of $DF^n(x_0)$ with $|\lambda| = 1$): the n-cycle is *asymptotically stable* (an *attractive* n-cycle) if all $|\lambda| < 1$; it is *unstable* (a *repelling* or *saddle* n-cycle) if some $|\lambda| > 1$. For the notion of convergence of an orbit to an attractive periodic orbit γ: $F^n(x_0) \to \gamma$, $n \to +\infty$, see the remarks in section 2.8.

3.8 Bifurcation of Maps

The motivating discussion of subsection 2.7.1 for flows applies here too. So it is clear that the analogue of the definition given there is: there is a *local bifurcation* of the map F_μ at the fixed point x^* when an eigenvalue λ of $DF_\mu(x^*)$ crosses the unit circle in the complex eigenvalue plane. If it crosses at the *bifurcation value* μ_b of the parameter, then $|\lambda(\mu_b)| = 1$ and this value is isolated. The same statement applies for local bifurcations near periodic points with the substitution of F_μ^n for F_μ.

The definition of *structural stability* and of a general *bifurcation* for a map is the strict analogue of (2-23), with the cascade $x_{n+1} = F_\mu(x_n)$ replacing the flow and the perturbations $\delta_1 F(x)$, $\delta_2 F(x)$ of $F_{\mu_b}(x)$ replacing the perturbations of $f_{\mu_b}(x)$.

> **Ex. 3.3** Find fixed points, their stability, and bifurcation values (if any) for the following real cascades. Draw the bifurcation diagrams.
>
> a) $F_\mu(x) \equiv \mu - x^2, \quad \mu \in \Re.$
>
> b) $F_\mu(x) \equiv \mu x(1-x), \quad 0 < \mu < 4.$
>
> c) $(x, y) \mapsto (y, -x/2 + \mu y - y^3), \quad \mu > 0.$ This is a cascade in $\Re^2$.

3.9 One-Dimensional Maps

For 1D maps there are special techniques available: we mention in particular the graphical technique for plotting the phase portraits. In the xy plane draw the graph of the curve $y = F(x)$ and the diagonal line $d : y = x$. Then an algorithm for constructing the orbit is to draw the steps bounded by the line d and the graph $y = F(x)$ as illustrated in Fig. 3.3. The projections of the points at which the steps meet the graph onto the x-axis then obviously give the orbit $x, F(x), F^2(x), \cdots$.

The intersections of the graph and the diagonal line d give the fixed points of the map. If F^n instead of F is plotted, these intersection points give the fixed points of order n, that is, the points of the n-cycles. Fig. 3.3 shows the behavior of orbits near attractive or repelling fixed points. Specializing the criterion of section 3.4 to $N = 1$, we have $\lambda \equiv dF(x^*)/dx$; and x^* is attracting if $|\lambda| < 1$, repelling if $|\lambda| > 1$. But since λ is geometrically the *slope* of the graph at x^*, we can determine the stability of a fixed point (or of an n-cycle) visually by noting whether the slope of the graph at the fixed point (or fixed point of F^n) is less than or greater than unity in magnitude. This can be seen in the figure, where the influence of Sgn λ on the orbit is also evident. Also shown is the case $\lambda = 0$, where for this particular map the orbit actually converges to x^*, though, as we remember, linear stability analysis is powerless in this case.

3.10 The Logistic Map

We illustrate some of these topics, in particular bifurcation, on a famous and well-studied 1D map

$$F_\mu(x) \equiv \mu x(1-x), \quad 0 \le x \le 1, \quad 0 \le \mu \le 4\,. \tag{3-9}$$

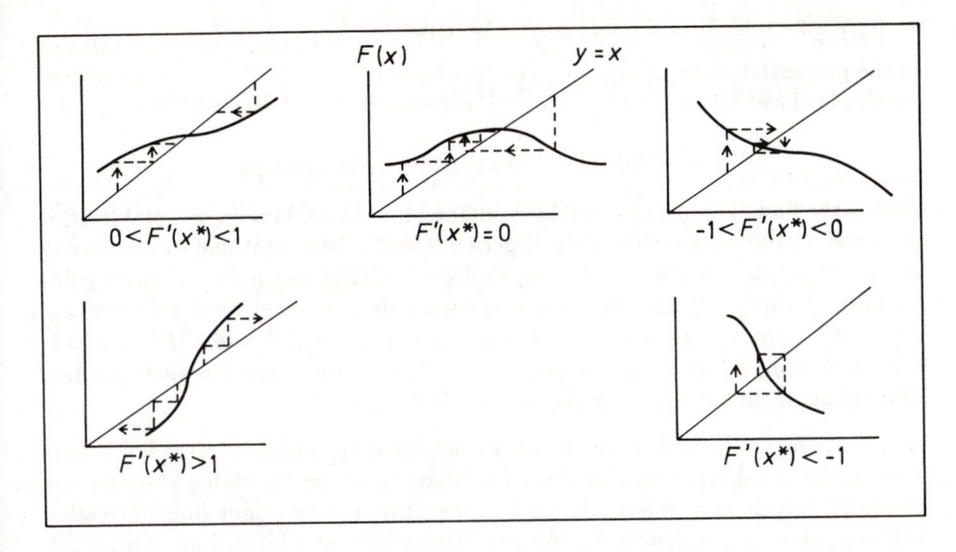

Fig. 3.3 Graphical method illustrated. Top: attractive fixed points; bottom: repelling fixed points.

See Bai-Lin [5] (section I4), Feigenbaum [16]. For the shown parameter range F maps the unit interval into itself, so that we can restrict our phase space to $0 \leq x \leq 1$. It is marvelous that such a simple-looking dynamics (one-dimensional, only a quadratic map!) has revealed an incredibly rich pattern of stability regimes, including "chaos," various parts of which show up in real nonlinear physical systems of the greatest complexity.

3.10.1 Period-Doubling

The fixed points of (3-9) are $x_1^* = 0$, $x_2^* = (\mu - 1)/\mu$. The eigenvalue $\lambda = F'_\mu(x^*) = \mu(1 - 2x^*)$, so $\lambda_1 = \mu$, $\lambda_2 = 2 - \mu$. Hence, in $0 < \mu < 1$, x_1^* is stable, $x_2^* \notin$ phase space. (In section 3.10 let "stable" be short for asymptotically stable, since we shall discuss the stability type only for hyperbolic regimes.) In $1 < \mu < 3$, x_1^* is unstable and x_2^* is stable. Thus $\mu_1 = 1$ is the first (local) bifurcation value. That is, at $\mu_1 = 1$ both $|\lambda_1|$ and $|\lambda_2|$ passed through 1; x_1^* lost stability and x_2^* gained stability.

At $\mu = 3$, $x_2^* = 2/3$ has $\lambda_2 = -1$, so x_2^* becomes unstable as μ increases through 3. So $\mu_2 = 3$ is the second (local) bifurcation value. Since there are no further fixed points of F_μ, what happens? You can check that F_μ^2 develops two *new* real fixed points $x_0, x_1, \in$ phase space, that is, which are not fixed points of F_μ. For $\mu \approx 3.04$, slightly greater than $\mu_2 = 3$, $x_0 = 0.5984\ldots$ and $x_1 \approx 0.7306\ldots$. Thus we have a 2-cycle. It is a *stable* 2-cycle because

$$|F_\mu^{2\prime}(x_0)| = |F_\mu^{2\prime}(x_1)| = |F_\mu'(x_0)F_\mu'(x_1)| = \mu^2|(1-2x_0)(1-2x_1)|$$

$$\approx \mu^2 \times 0.0908 < 1 \text{ in } \mu_2 < \mu < \mu_3 \text{ for some } \mu_3.$$

Here is the first example of a period-doubling bifurcation, here from a stable orbit of period $1 = 2^0$ to a stable orbit of period $2 = 2^1$. Now continue to increase μ; at $\mu_3 = 1 + \sqrt{6} \approx 3.4495\ldots$ the 2-cycle loses stability and a $4 = 2^2$-cycle gains stability. Namely, F_μ^4 develops four *new* fixed points y_j – not fixed points of F_μ, F_μ^2, or F_μ^3 – and these are stable: $|F_\mu^{4\prime}(y_j)| < 1$, $j = 0,1,2,3$, while $|F_\mu^{2\prime}(x_i)| > 1$, $i = 0,1$, in $\mu_3 < \mu < \mu_4$ for some μ_4. This is the second period-doubling bifurcation, from a stable 2^1-cycle to a stable 2^2-cycle.

As μ increases still further, there is a *period-doubling cascade* ("cascade" is here used in the usual sense, not as short for "map dynamics."): stable 2^n-cycle $\longrightarrow$ stable 2^{n+1}-cycle as $n \to +\infty$. It can be shown that the corresponding bifurcation values $\mu_{n+2} \to \mu_\infty \approx 3.5699\ldots$. We can display this on a bifurcation diagram in which only the stable 2^n-cycles are shown (Fig. 3.4).

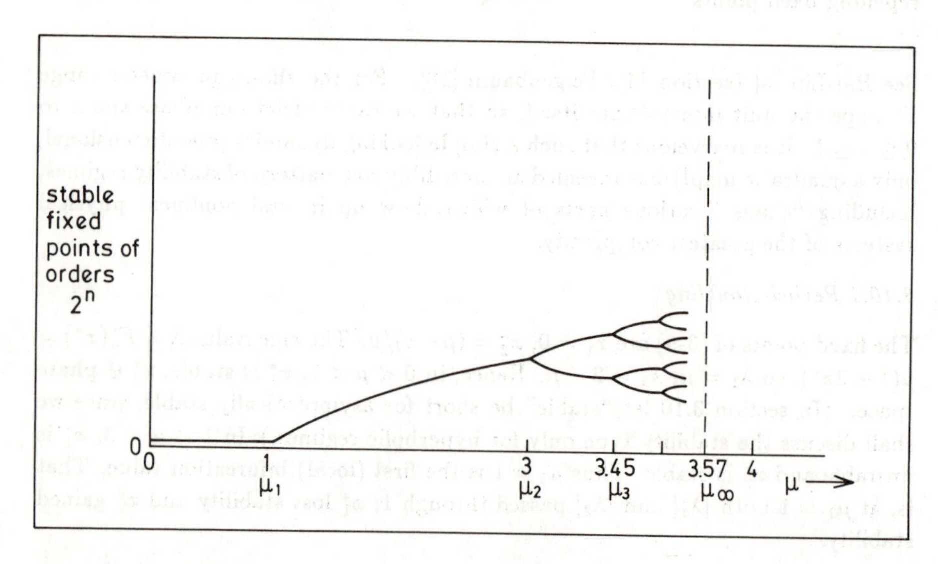

Fig. 3.4 Partial bifurcation diagram for the logistic map (not to scale)

Fig. 3.4 ends at μ_∞. We postpone the complete bifurcation diagram until we discuss a few more topics.

Consider now the ratio of successive parameter intervals between period-doubling bifurcations. The limiting ratio

$$\delta \equiv \lim_{n\to\infty} \frac{\mu_n - \mu_{n-1}}{\mu_{n+1} - \mu_n} \approx 4.6692016091\cdots \qquad (3-10)$$

does exist, so these parameter intervals decrease asymptotically as a geometric sequence. The ratio δ is a *universal number* in the sense that it turns up in many, diverse physical phenomena which show period-doubling behavior. See, for example, Chapter 8. Since the convergence is rapid, (3-10) can be used to predict μ_{n+1}, given μ_n and μ_{n-1}, with fair accuracy. For example,

$$\mu_4 \approx \mu_3 + \delta^{-1}(\mu_3 - \mu_2) \approx 3.45 + (3.45 - 3)/4.67 \approx 3.55,$$

which is correct to two decimal places.

3.10.2 Graphical Treatment of Period-Doubling

The graphical method of section 3.9 can be applied to elucidate the fundamental geometrical reason for the period-doubling phenomenon. This adumbrates the ideas of self-similarity, renormalization group, and fractals.

Consider first Fig. 3.5. We can literally see how the stability of the fixed points of F_μ and F_μ^2 changes as μ is increased from 0 to 3.

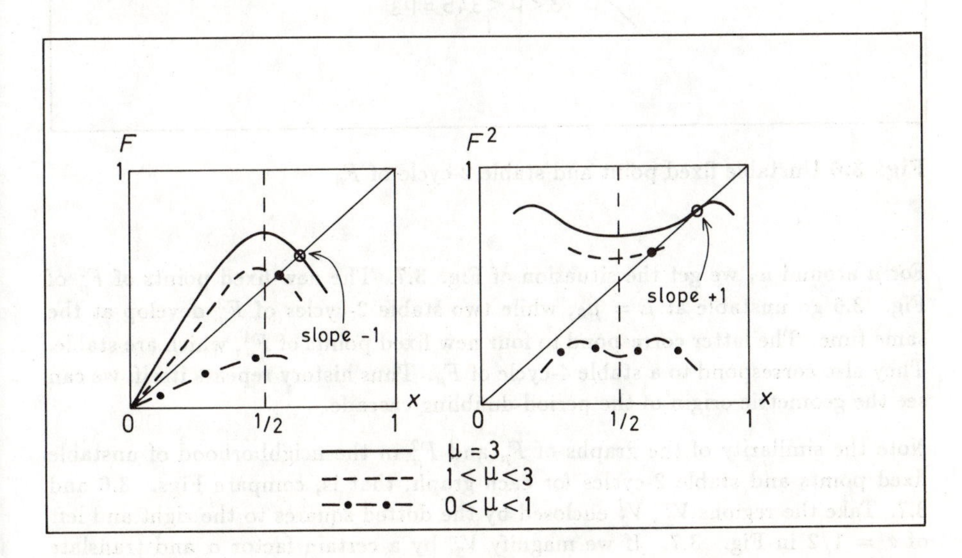

Fig. 3.5 Bifurcations of the logistic map by the graphical method

The behavior described in subsection 3.10.1 can be read off from the slopes of the respective functions at their intersections with the diagonal. The zero fixed point of F_μ is at first stable. At $\mu = 1$ it goes unstable while the new, nonzero fixed point gains stability for the range $1 < \mu < 3$. In the range $0 < \mu < 3$ these are the only fixed points of F_μ^2 in phase space, and they have the same stability for F_μ^2 as for F_μ. Fig. 3.6 shows what happens for $3 < \mu < \mu_3 \approx 3.45$. The iterated map F_μ^2 suddenly gains two new fixed points at $\mu = 3$. They are stable, and correspond to the stable 2-cycle of F_μ shown. The steps, constructed by the graphical algorithm, close on themselves, the mark of a periodic orbit.

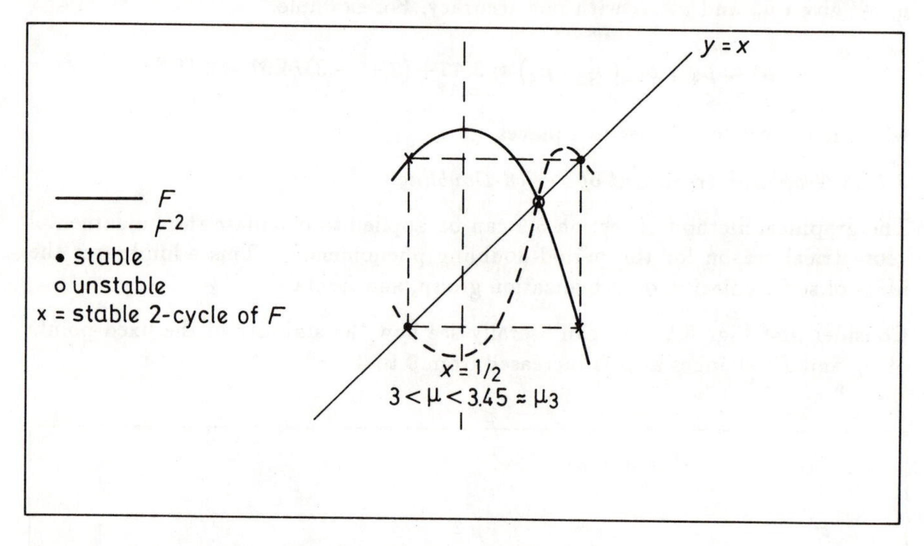

Fig. 3.6 Unstable fixed point and stable 2-cycle of F_μ

For μ around μ_3 we get the situation of Fig. 3.7. The new fixed points of F_μ^2 of Fig. 3.6 go unstable at $\mu = \mu_3$, while two stable 2-cycles of F_μ^2 develop at the same time. The latter correspond to four new fixed points of F_μ^4, which are stable. They also correspond to a stable 4-cycle of F_μ. Thus history repeats itself; we can see the geometric origin of the period-doubling cascade.

Note the similarity of the graphs of F_μ and F_μ^2 in the neighborhood of unstable fixed points and stable 2-cycles for each graph, that is, compare Figs. 3.6 and 3.7. Take the regions V_2^r, V_2^l enclosed by the dotted squares to the right and left of $x = 1/2$ in Fig. 3.7. If we magnify V_2^r by a certain factor α and translate it to coincide with the region V_1 in the dotted square of Fig. 3.6, the graphs of $F_{\tilde{\mu}}^2$ and F_μ practically coincide for the appropriate μ and $\tilde{\mu}$. The same is

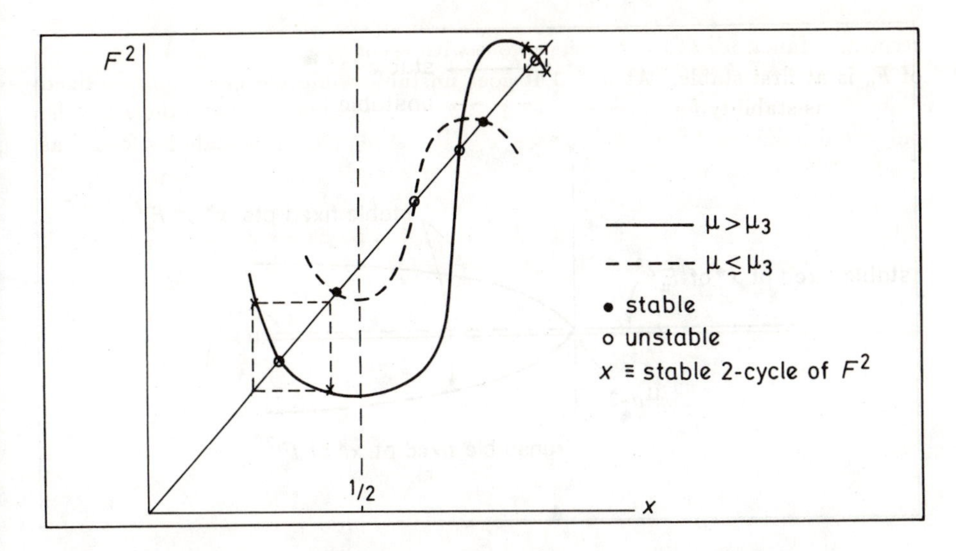

Fig. 3.7 Bifurcation of the iterated logistic map F_μ^2 by the graphical method

true for V_2^l after an inversion in the point $(x,y) = (1/2,1/2)$. And similarly for all further period-doublings. The magnification factor α quickly approaches another universal pure number. Thus a graph displaying *all* $F_{\tilde{\mu}_n}^{2^n}$, $n = 0,1,2,3,\cdots$, together, for appropriate parameter values $\tilde{\mu}_n$ in the ranges $\mu_{n+1} < \tilde{\mu}_n < \mu_{n+2}$, shows *self-similarity on all length scales.* Such a set is called a *fractal.*

3.10.3 Tangent Bifurcations

We have seen that the slope of $F_\mu^{2^n}$ at one of its fixed points decreases through -1 at $\mu = \mu_{n+2}$, a period-doubling bifurcation. This is a pitchfork bifurcation for $F_\mu^{2^n}$ (Fig. 3.8).

There is another very important local bifurcation for the logistic map, the tangent bifurcation, which leads to intermittent behavior. *Intermittency* is defined generally as regular behavior interrupted by random bursts of "chaotic" behavior. We can illustrate it on the simple 1D map

$$G_\mu(x) \equiv \mu + x - x^2$$

for x and μ around zero (Fig. 3.9).

For $\mu \lesssim 0$, there is a narrow neck between the graph of G_μ and the diagonal line. The orbit point spends a long time in this neck and is apparently converging to the false fixed point $x = 0$. Once out of the neck, there may be "chaotic" behavior in the general system which shows intermittency until the orbit point falls "by

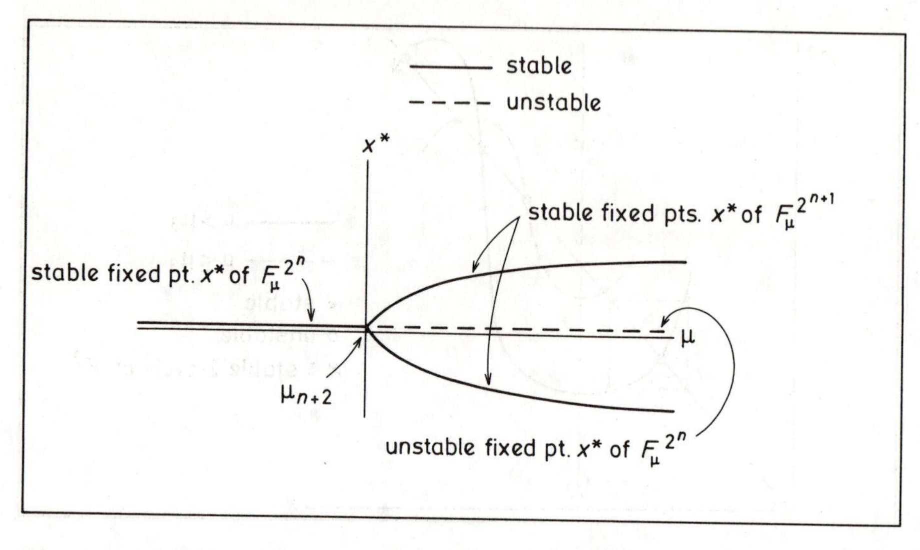

Fig. 3.8 Period-doubling bifurcation for the logistic map: a pitchfork bifurcation of $F_\mu^{2^n}$

chance" into the entrance to the neck again, and the regular behavior repeats. As μ increases through 0, the bifurcation value for the model G_μ, the intermittent regime disappears, and a pair of fixed points, one stable, one unstable, is born (Fig. 3.10).

This happens in the parameter regime $\mu_\infty < \mu < 4$ for the logistic map (3-9); pairs of m-cycles of *odd* order $m = k2^n$, k = odd integer, are born, preceded by intermittency. Thus we are saying that for μ around one of these tangent bifurcation values μ_b, the dynamics of the logistic map F_μ or one of its iterates F_μ^m is topologically equivalent to the dynamics of the model map G_μ, $\mu \approx 0$, in sufficiently small neighborhoods X or X_m of phase space $[0, 1]$.

The full bifurcation diagram of the logistic map in $0 < \mu \leq 4$ is shown in Fig. 3.11.

We have explored only part of the stability structure of the regime $(\mu_\infty, 4]$ here; for more details see the cited specialized works.

However, one further feature of this regime should be mentioned here: it is the "chaotic" regime in the sense that the Lyapunov exponent $\chi > 0$. (We have already seen this for the single value $\mu = 4$ in section 1.3.)

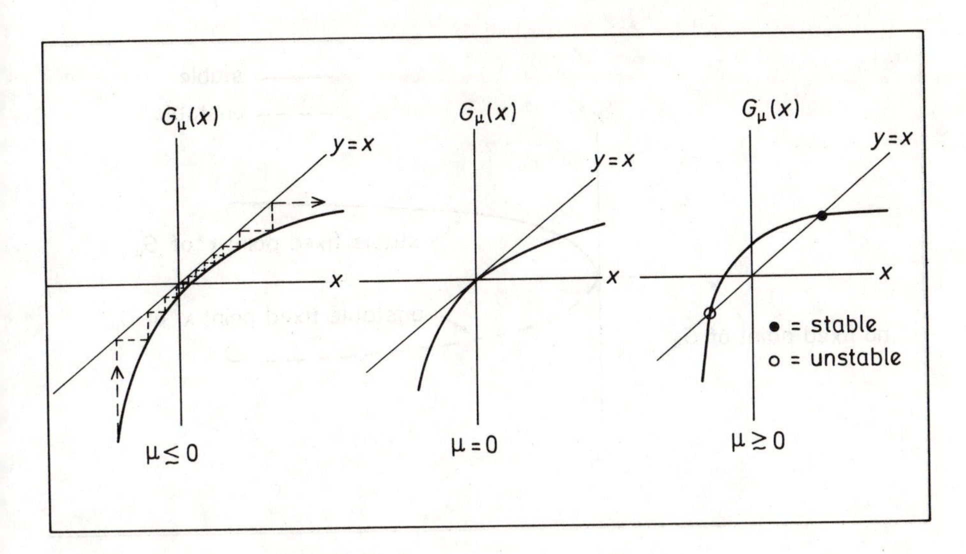

Fig. 3.9 Tangent bifurcation by the graphical method

3.10.4 Lyapunov Exponent for 1D Maps

Let $x_n = F^n(x_0)$ be an orbit of a 1D map F. For an initially nearby orbit we have $x'_0 = x_0 + \epsilon$, ϵ small. Then

$$\epsilon \equiv x'_n - x_n \approx F^{n\prime}(x_0)\epsilon = \epsilon \prod_{j=0}^{n-1} F'(x_j)$$

by the chain rule (3-3). Take absolute value, write $|\epsilon_n| \equiv e^{n\chi_n}|\epsilon|$, take the natural logarithm, and divide by n. This defines χ_n. Then we have the definition

$$\chi \equiv \lim_{n\to\infty} \chi_n = \lim_{n\to\infty} \frac{1}{n} \sum_{j=0}^{n-1} \ln|F'(x_j)|. \qquad (3-11)$$

If this limit exists, χ is called the *Lyapunov exponent* (LE) of the orbit based at x_0. This can be generalized to maps and flows in $\Re^N$ and even beyond phase spaces $\subset \Re^N$, but the generalization is not trivial. We postpone this to Chap. 5.

You see that χ is the average of the linearization growth exponent $\ln|F'(x)|$ along the orbit. In fact, if x_0 is a fixed point x^*, then $\chi = \ln|\lambda| \equiv \ln|F'(x^*)|$, and is > 0, $= 0$, or < 0 according as $|\lambda| > 1$, $= 1$, or < 1. If $\chi > 0$ for a general orbit, then it is diverging (*initially*!) at an exponential rate from any orbit which is sufficiently close to it at some time n. This is a popular signature of sensitive dependence

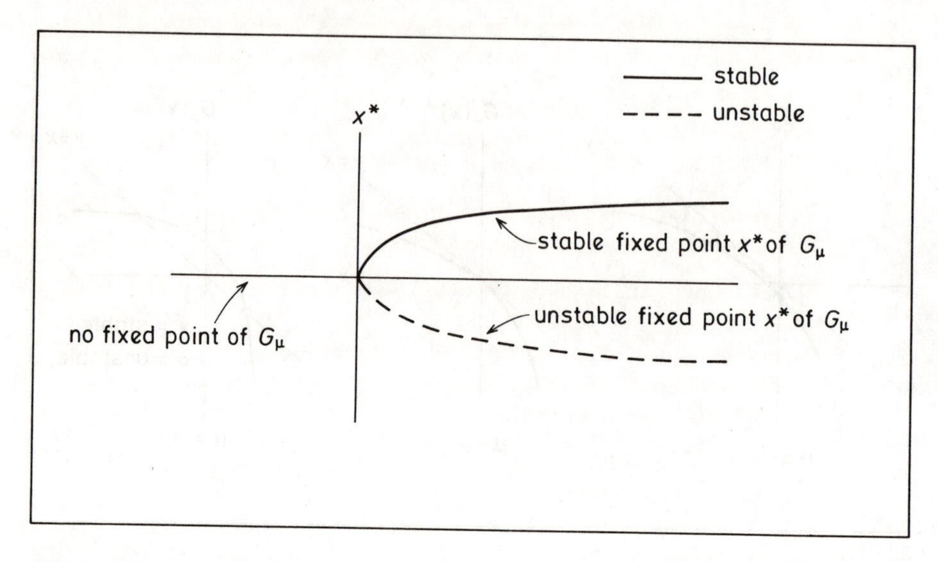

Fig. 3.10 Tangent bifurcation diagram

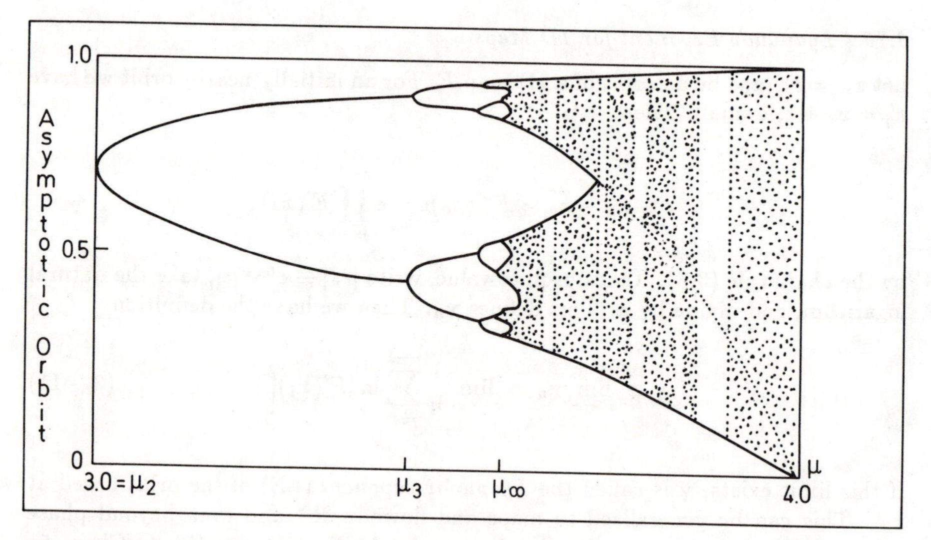

Fig. 3.11 Full bifurcation diagram for the logistic map

(SD) in the physics literature. Note also that the LE of a stable n-cycle < 0 (Ex. 3.4).

Ex. 3.4 Prove that $\chi < 0$ for an (asymptotically) stable fixed point or periodic orbit.

Prob. 3.1 For the logistic map (3-9) and specific parameter values $\mu = 1.5$, 3.1, 3.5, 3.5699, 3.6, 3.828, and 4, use a microcomputer or programmable calculator to do the following:

a) Plot the orbit for each μ for a typical initial value x_0, $0 < x_0 < 1$, and a sufficiently large number n of iterates.

b) Try to guess the stability types of whatever attractors, if any, exist. (Hint: it is essential that you *plot* the orbits, not just list the points in a computer printout, so as to be able to identify the asymptotic [large n, after transients die out] behavior "by eye.") For example, "stable 3-cycle" might be an answer here.

c) Calculate the LE χ of each of these orbits, and try to correlate χ with the observed stability type.

3.11 Poincaré Maps

A Poincaré map of a flow is one of the most valuable tools for studying flows in nonlinear dynamics. It reduces the dimension of the phase space by at least 1; it reduces continuous orbits to discrete orbits (easier to compute); nevertheless, it gives a complete and equivalent picture of the stability or "chaotic" behavior in sufficiently small neighborhoods.

A Poincaré map is commonly used to study flows in three situations:

- To study the stability of periodic orbits (limit cycles).
- To study time-periodic flows. These are nonautonomous flows such that $f(x,t) = f(x,t+T)$ for some T, all $x \in \Re^N$. These usually arise from periodic forcing terms in motion equations.
- To study the flow near a *homoclinic* or *heteroclinic orbit.* These are orbits which join a hyperbolic fixed point x^* to itself, or which join two hyperbolic fixed points x_1^*, x_2^*, respectively. Such an orbit must therefore lie in the intersection $W^s \cap W^u$, where both manifolds belong to x^* in the homoclinic case, or one belongs to x_1^*, the other to x_2^*, in the heteroclinic case. Such orbits are not usually structurally stable, so any slight change of the parameter breaks them in two. "Chaotic" flow then ensues.

The subject of Poincaré maps is a huge, mathematically deep one. We will only sketch the key idea here, and assign an illuminating problem. For the theory, see GH (section 1.5) or Wiggins [36] (section 1.2) for the basics. Most of their succeeding chapters use Poincaré maps as an essential tool.

Consider the application (1) above. Let a flow $\dot{x} = f(x)$ in $\Re^N$ with orbit function now written $\phi(t, x)$ have a periodic orbit of period T. We pass an $(N-1)$-dimensional surface Σ through the flow, cutting γ transversely at x_0. Take a smaller open set $V \subset \Sigma$ which contains x_0. Then if V and Σ are small enough, the orbits which pass through V will intersect Σ when they come around again, in a time close to T. The map that associates points in V with their *points of first return* in Σ is called the *Poincaré map*; call it P. Then more precisely,

$$P : V \longrightarrow \Sigma \quad \text{by} \quad x \longmapsto \phi(\tau(x), x), \qquad (3-12)$$

where $\tau(x)$ is the time of the first return of x to Σ. Note $\tau(x_0) = T$ and $\phi(T, x_0) = x_0$. Thus P maps x_0 into x_0; x_0 is a fixed point of the Poincaré map. Σ is called a *cross section* of the flow f. Figure 3.12 will make all of this clear.

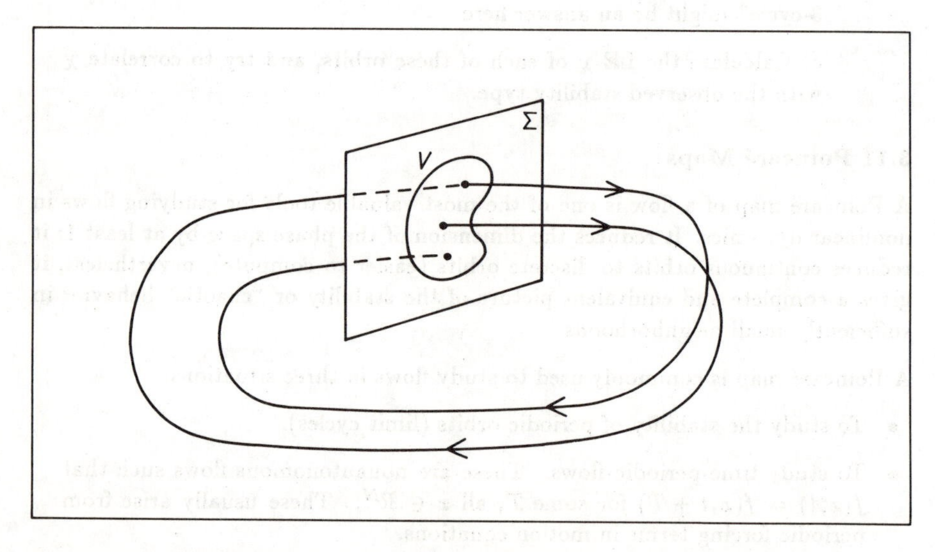

Fig. 3.12 Illustrating the Poincaré map

Practically speaking, one must have a pretty good idea of the flow phase portrait before being able to choose a cross section intelligently. However, the choice of a particular Σ satisfying the definition above is not too critical, because any two such maps, P and P', are topologically equivalent!

Now consider application (2) above. The nonautonomous flow is $\dot{x} = f(x, t)$, and obeys $f(x, t+T) = f(x, t)$ for all x. Let $\phi_t(x_0)$ be the orbit function, with $\phi_0(x_0) = x_0$. Then define the map P by

$$P(x_0) \equiv \phi_T(x_0), \quad \text{all } x_0 \in \Re^N. \qquad (3-13)$$

That is, we simply sample the flow at times nT, $n = 0, \pm 1, \pm 2, \pm 3, \cdots$, to define the map orbit. For note $P^n(x_0) = \phi_{nT}(x_0)$. (Of course, this map can be used for autonomous flows as well.) Unlike case (1), P need not have a fixed point. If it does, say $P(x^*) = x^*$, then x^* corresponds to a periodic orbit of period T for the flow. A fixed point of order $n : P^n(x^*) = x^*$ corresponds to a *subharmonic* of the flow of period nT.

The map (3-13) is really of the Poincaré type, as you can see by making the flow autonomous by writing $\dot{x} = f(x, \theta), \dot{\theta} = 1$; then the cross section Σ is the N-dimensional surface $\theta = 0$ cutting the flow transversely in the enlarged phase space $\Re^N \times S^1$, where S^1 is the unit circle (that is, θ is an angle and measured mod 2π). Then the time of first return $\tau(x) = T$ is the same for all $x \in \Re^N = V = \Sigma$.

Prob. 3.2 Consider the harmonically forced, damped, *Duffing oscillator*

$$\ddot{x} + \delta\dot{x} - x + x^3 = \gamma \cos \omega t, \quad \delta > 0. \qquad (3-14)$$

This problem is to be done using a programmable computer with a graphics display. Take δ and ω fixed, say $\delta = 0$ or 0.25 and $\omega = 1$, and consider γ variable, $0 \leq \gamma < 1$.

a. Unforced case $\gamma = 0$. Put (3-14) into autonomous flow form in $\Re^2$. Find the fixed points, their invariant subspaces E^u, E^s, E^c, plot a phase portrait which indicates also the invariant manifolds W^s, W^u of each fixed point. Do this for $\delta = 0$ and $.25$.

b. Now let γ be nonzero: consider the cases $\gamma = 0.20, 0.30, 0.40$, all with $\delta = .25$. Use the Poincaré map $P_\gamma \equiv (3-13)$ with $T \equiv 2\pi/\omega = 2\pi$. Numerically plot W^s and W^u for the saddle fixed point of P_γ near $(0,0)$.

Remarks and hints on Prob. 3.2: This is a fairly ambitious problem. Part (a) can be done more or less analytically with good intuition. As to part (b), luckily the map P_γ is *structurally stable* (section 3.8) at $\gamma = 0$, and so it continues to have three hyperbolic fixed points near those of part (a) for small γ. To plot W^s and W^u numerically for the saddle point x^*, determine the directions E^s and E^u and then iterate a number of initial points on these directions close to x^*. (Remember that W^s and E^s, W^u and E^u are *tangent* at the fixed point by the Stable Manifold theorem, see section 3.6!) To get W^s, iterate P_γ^{-1}; to get W^u, iterate P_γ. There is a (global) bifurcation somewhere between $\gamma = 0.10$ and 0.20. You will find a *transverse homoclinic orbit* (section 3.6) developing when W^s and W^u intersect transversely, thus "chaotic" motion. For γ large, ≈ 0.40 or greater, the graphics should reveal what appears to be a strange *attractor* (see section 2.8; an obvious analogous definition holds for maps). It apparently coincides with W^u. You may want to compare your results with GH, section 2.2.

4. HAMILTONIAN SYSTEMS

4.1 Generalities

The flow

$$\dot{q} = \partial H/\partial p, \quad \dot{p} = -\partial H/\partial q, \quad q, p \in \Re^n, \tag{4-1}$$

in $\Re^N$, $N \equiv 2n$, is a *Hamiltonian system of n degrees of freedom.* $H \equiv H(q,p,t)$ is called the *Hamiltonian.* If H does not explicitly depend on t, the flow is autonomous, and $H(q(t),p(t))$ is a *constant* (or *integral*) *of the motion.* In other words, H is *conserved.* For we can easily prove, using the motion Eqs. (4-1), that

$$dH/dt = \partial H/\partial t, \tag{4-2}$$

which equals zero in the autonomous case. Then the flow is confined to the $(2n-1)$-dimensional *energy surfaces* $H(q,p) =$ const. in phase space.

Hamiltonian flows *conserve volume in phase space*, and are nondissipative, since $\nabla \cdot f = 0$. Let us check this:

$$\nabla \cdot f \equiv \sum_{k=1}^{2n} \partial f_k/\partial x_k \equiv \sum_{i=1}^{n} \left[\frac{\partial}{\partial q_i}\left(\frac{\partial H}{\partial p_i}\right) + \frac{\partial}{\partial p_i}\left(\frac{-\partial H}{\partial q_i}\right)\right] = 0, \tag{4-3}$$

whether or not H depends on t. Hence they can have no low-dimensional attractors of the dissipative type, where an initial nonzero volume in phase space shrinks in time to zero volume (section 2.8). Hamiltonian flows are usually considered more fundamental than dissipative flows, since microscopic physical motion equations like (4-1) possess time-reversal invariance while dissipative flows do not. (Note the dissipative term $+\delta\dot{x}$ in the damped Duffing oscillator (3-14), for example.) Dissipative systems in $\Re^N$ are truncated models of T-invariant systems in a larger phase space $\Re^{N'}$, $N' >> N$, where the interaction between parts of the larger system is modelled as dissipative terms in a smaller system.

As for the question of stability, note that all *stable* fixed points of (4-1) must be *centers* (the asymptotically stable, or sink, case is excluded). Refer to section 2.3

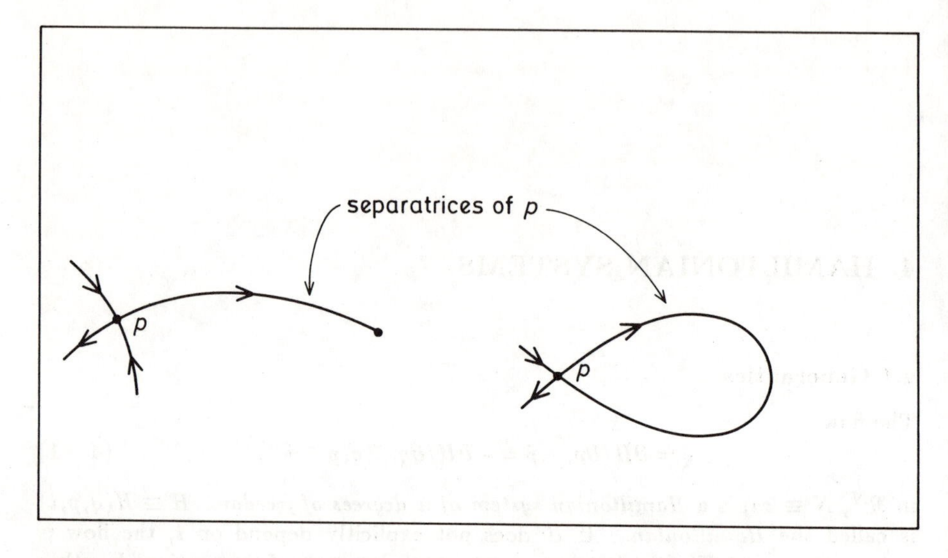

Fig. 4.1 Two cases of separatrices

to recall these definitions. We can see this as follows. Consider a stable fixed point x^*.

$$0 = \nabla \cdot f\Big|_{x^*} \equiv \Big(\sum_k \partial f_k/\partial x_k\Big)\Big|_{x^*} \equiv Tr \;\; Df(x^*) =$$

$$= \sum_k \lambda_k \Rightarrow \sum_{k=1}^{N} Re\lambda_k = 0. \qquad (4-4)$$

Here the $\{\lambda_k\}$ are the eigenvalues of the Jacobian matrix at the fixed point. But if x^* is stable, no $Re\lambda_k > 0$ (this seems evident, and can be proved). Therefore no $Re\lambda_k < 0$ from (4-4), since the real parts must add to zero! Thus *all* $Re\lambda_k = 0$, $k = 1, 2, \cdots N$. This says that $E^u = E^s = \mathbf{0}$; only $E^c \neq \mathbf{0}$. Then it can further be proved that this guarantees that x^* is in fact a center for most ("typical," "nonpathological") Hamiltonians. One says: "under *generic* conditions." So from now on we assume that all fixed points of a Hamiltonian system are either centers or saddles.

Examples. A simple, important, autonomous Hamiltonian system for $n = 1$ is the linear harmonic oscillator $H(q,p) = (p^2 + q^2)/2$, in properly scaled variables. There is one fixed point $(0,0)$, a center, and all orbits are periodic with the same period $\tau \equiv 2\pi/\omega$, where $\omega \equiv \sqrt{k/m} = 1$ here.

A *separatrix* is a branch of $W^u(p)$ or $W^s(p)$ for a hyperbolic fixed point p (Fig. 4.1).

The general definition of *hyperbolic* fixed point was given in section 2.4; for a Hamiltonian system there is only one possibility: p is a *saddle* – the source and sink cases are out. The harmonic oscillator has no separatrices since it has no saddles. However, another, all-important example of an $n = 1$ Hamiltonian system does have separatrices. This is the 1D *pendulum* (Fig. 4.2):

$$\dot{\theta} = p, \quad \dot{p} = -\alpha \sin\theta \; ; H(\theta, p) = p^2/2 - \alpha\cos\theta. \tag{4-5}$$

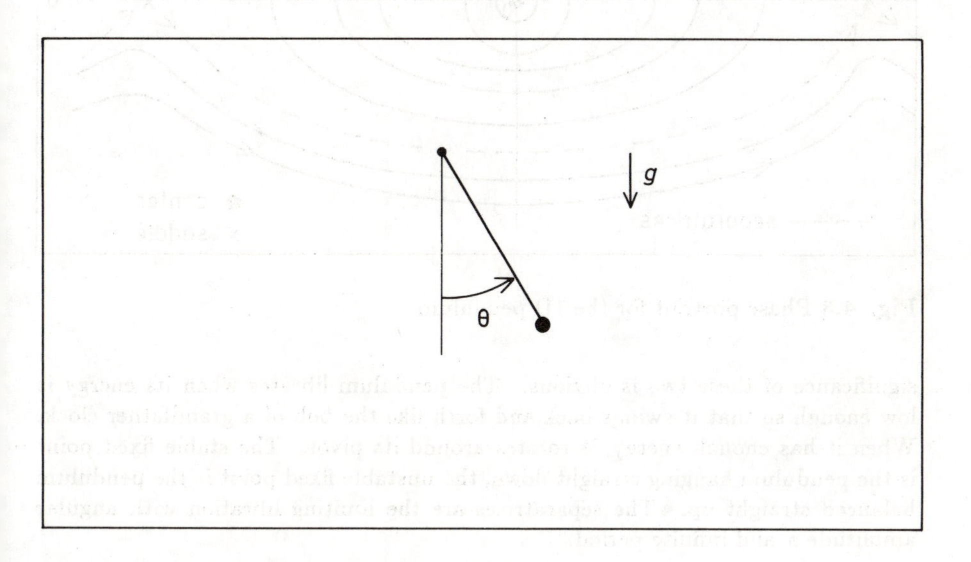

Fig. 4.2 The 1D pendulum

Phase space here is actually $\Re \times S^1$, $S^1 \equiv$ unit circle, since θ is an angle, measured mod 2π. Thus, topologically it is the surface of a cylinder rather than the plane $\Re^2$. The *different* fixed points are two, $(0,0)$, a center, and $(\pi, 0)$, a saddle. For $(-\pi, 0)$ is to be identified with $(\pi, 0)$, and (π, π) with $(0,0)$, etc. The phase portrait looks like Fig. 4.3.

Note the two separatrices, so-called because they separate regions of qualitatively different orbits. Inside them the orbits are closed, going from approximately circular harmonic oscillator orbits near the center to distorted limit cycles whose periods $\rightarrow \infty$ as they approach the separatrices. (Why is this obvious?) This type of periodic motion is called *libration* (*def.* the orbit is closed in phase space). Outside the separatrices the orbits are open curves. This type of motion is called *rotation* (*def.* $p(q)$ is a periodic function). So the harmonic oscillator shows only libration, while the pendulum shows both libration and rotation. The geometric

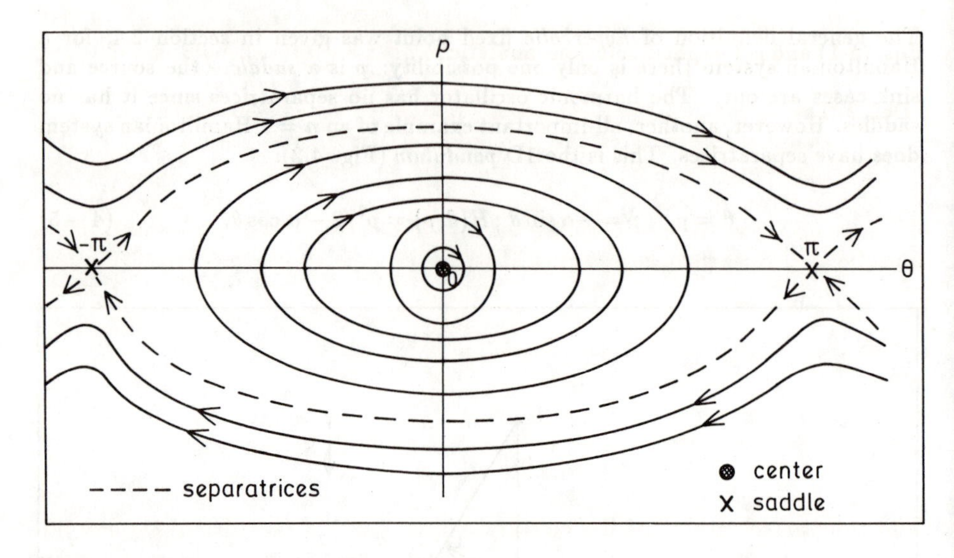

Fig. 4.3 Phase portrait for the 1D pendulum

significance of these two is obvious. The pendulum librates when its energy is low enough so that it swings back and forth like the bob of a grandfather clock. When it has enough energy, it rotates around its pivot. The stable fixed point is the pendulum hanging straight down, the unstable fixed point is the pendulum balanced straight up. The separatrices are the limiting libration with angular amplitude π and infinite period.

For a good treatment of Hamiltonian systems see Goldstein [18].

4.2 Integrable Systems

There are three presumably equivalent definitions of an integrable Hamiltonian system. We give all three, since each one casts a different light on the concept. From now on we assume an autonomous system with analytic Hamiltonian $H(q,p)$. For real-valued functions f of real variables $x \in \Re^N$, we have the

Def. f is *analytic* in a neighborhood $V \subset \Re^N$ if $f(x)$ admits a power series expansion in x about every point $x_0 \in V$. $\quad (4-6)$

Thus we are assuming that H has a power series expansion in $x = (q,p) \in \Re^{2n}$ except possibly at some singularities.

Def. 1 The system (4-1) is *integrable* if the solution curves $\phi_t(x_0) \equiv (q(q_0,p_0,t), p(q_0,p_0,t))$ are analytic in $x_0 = (q_0,p_0)$ and in t. $\quad (4-7)$

This is the most fundamental definition. One can simply *integrate* the differential Eqs. (4-1) and get explicit solutions analytic in both initial conditions and in time.

Def. 2 The system (4-1) is *integrable* if there exist n independent analytic constants of the motion $F_i(q(t), p(t)) = \text{const.}$, $i = 1, 2, \cdots n$. $(4-8)$

There is already one constant of the motion, H itself, (see (4-2)). However, there must be $n-1$ further conserved quantities, and all *independent* (no F_i is a function of the other F_j, $j \neq i$) if the system is to be integrable. These usually arise when the system is not coupled to external fields, which destroy the symmetry of H under translations or rotations. Then one gets total linear momentum or total angular momentum, etc., conserved as well as the energy.

Def. 3 The system (4-1) is *integrable* if its Hamilton-Jacobi equation admits a complete solution. $(4-9)$

This is more technical. The H-J equation is the single partial differential equation

$$H(q, \partial W/\partial q) = E = \text{const.}, \tag{4-10}$$

for Hamilton's characteristic function W. A *complete solution* of (4-10), $W(q_1, q_2, \cdots q_n\ ;\ \gamma_1, \gamma_2, \cdots \gamma_n)$, admits n *independent* integration constants γ_i. (The energy E can be chosen to be one of these.) We explain briefly how Eq. (4-10) arises, and refer the reader to Goldstein [18] for details.

Consider an analytical *canonical transformation* $Q_i = Q_i(q, p)$, $P_i = P_i(q, p)$, $i = 1, 2, \cdots n$, with new Hamiltonian defined by $K(Q, P) \equiv H(q, p)$, that is, such that the Hamiltonian form (4-1) is preserved:

$$\dot{Q} = \partial K/\partial P, \quad \dot{P} = -\partial K/\partial Q. \tag{4-11}$$

What if we could find a canonical transformation such that K did not involve the $Q_i : K = K(P)$. Then we could integrate (4-11) *immediately and trivially*, getting

$$P_i(t) = \gamma_i = \text{consts.}, \quad Q_i(t) = \alpha_i t + \beta_i\ ;\ \alpha_i, \beta_i \text{ consts.}, \tag{4-12}$$

where $\alpha_i(\gamma_i) \equiv (\partial K/\partial P_i)\Big|_{P_i = \gamma_i}$. How would we find such a canonical transformation? If it were generated by a function $W(q, P)$, we would have

$$p = \partial W/\partial q, \quad Q = \partial W/\partial P,$$

(see Goldstein [18]). We could get a partial differential equation for W by rewriting $H(q, p) = K(\gamma_i) = E = \text{const.}$ as in (4-10). Then it can be shown that a complete solution of (4-10) guarantees that the canonical transformation $(q,p) \mapsto (Q,P)$ can be inverted to get the solution

$$q_i = q_i(\alpha t + \beta, \gamma), \quad p_i = p_i(\alpha t + \beta, \gamma), \quad \alpha, \beta, \gamma \in \Re^n,$$

from (4-12). The $2n$ independent integration constants γ_i, β_i can be replaced by the $2n$ initial values q_{0i}, p_{0i}, and we are back to the explicit analytic solutions demanded by Def. 1, (4-7). This makes a long story short, but I hope that it gives the essential idea without oversimplifying.

You may well ask why we don't simply solve the H-J equation for the system of interest, thereby apparently getting explicit orbit solutions and proving any system integrable by Def. 1. The answer is, *a complete solution of the H-J equation does not always exist*; in fact, it is rare (a fact not sufficiently emphasized in our standard classical mechanics texts). In those cases the system is *nonintegrable.* A famous example is the Newtonian 3-body problem, which worried Newton and all his successors.

Some simple examples of integrable systems:

a. All 1D systems ($n = 1$). For, solve for $p = \phi(q)$ from $H(q,p) = E$. Substitute this into the motion equation $\dot{q} = \partial H/\partial p$, getting a first order ordinary differential equation for q. Integrate.

b. All linear flows. For, go to normal modes (that is, separate the variables), getting the equations $\ddot{q}_i + \omega_i^2 q_i = 0$, $i = 1, 2, \ldots n$. Integrate.

4.2.1 Action-Angle Variables

Let us consider a special case of an integrable system. First we assume that the H-J equation is *completely separable* in some set of canonical coordinates q, p. This means that it reduces to the n ordinary differential equations

$$H_i(q_i, \partial W_i/\partial q_i) = E_i = \text{consts.}, \quad \text{where}$$

$$H \equiv \sum_i H_i(q_i, p_i), \quad \sum_i E_i = E. \tag{4-13}$$

Then one gets the solution $W = \sum_i W_i(q_i; \alpha)$ [$\alpha \in \Re^n$ are some general set of n independent constants, functions of the separation constants E_i] by *quadratures* (solve the ith ordinary differential equation in (4-13) for $\partial W_i/\partial q_i$ as a function of q_i and E_i, and integrate). Second, we assume that (at least in some energy regimes) the motion is of *libration* or *rotation type.* This means, more precisely, that the projection of the motion on the q_i-p_i 2-plane for each $i = 1, 2, \cdots n$, is libration or rotation as defined for the 1D pendulum in section 4.1. We confine attention to these special integrable systems from now on.

Now we can introduce action-angle variables J, θ, a special case of the separable coordinates of (4-13). Define the *actions*

$$J_i \equiv \oint p_i dq_i \equiv \oint \frac{\partial W}{\partial q_i}(q_i;\alpha)dq_i \equiv J_i(\alpha), \qquad (4-14)$$

where the integral $\oint$ is carried out over a complete period of the libration or rotation. Inverting (4-14), we can express the α_i as functions of $J \in \Re^n$. Since $K(\theta, J) \equiv H(q,p) = \sum H_i = \sum_i E_i(\alpha)$, K is a function $K(J)$ of the J only. Then the canonical motion Eqs. (4-11) have the simple solutions (4-12). Namely,

$$\dot{\theta}_i = \partial K/\partial J_i \equiv \nu_i(J) = \text{consts.} \quad \Rightarrow \theta_i(t) = \nu_i t + \beta_i$$

$$\dot{J}_i = -\partial K/\partial \theta_i = 0 \quad \Rightarrow J_i(t) = J_i = \text{consts.} \qquad (4-15)$$

Because of the definition (4-14) of the actions, these particular coordinates have special, nice properties. We just list them, and refer the interested reader to Goldstein ([18] Chap. 10) for proofs and details.

a. The $\nu_i \equiv \partial K(J)/\partial J_i$ are the frequencies of the libration or rotation.

b. The bounded motion, that is, libration, $x(t) = (q(t), p(t))$ is *quasi-periodic* with the frequencies ν_i:

$$x(t) = \sum_j a_j exp[2\pi i j \cdot (\nu t + \beta)], \quad j, \nu, \beta \text{ consts.} \in \Re^n. \qquad (4-16)$$

Here $j \cdot \nu \equiv \sum_i j_i \nu_i$ is the scalar product, j is an *integer vector*, namely, its n components are integers, a_j are a set of constant amplitudes in $\Re^{2n}$ indexed by integer vectors, and the sum in (4-16) goes over all integer vectors. We only mention that the unbounded motion (rotation) is similar, except that it has an extra term $q_0 \cdot (\nu t + \beta)$ linear in t in $q(t)$.

Property (a) has been a boon in celestial mechanics, since one can compute the fundamental frequencies just by finding $K(J)$, without solving for the complete motion. As for (b), note that the motion is in general *not* periodic, since the orbits do not close. Only in the case that the frequencies are *commensurable*,

$$j_r \cdot \nu = 0, \quad r = 1, 2, \cdots n-1 \qquad \text{(commensurable case)} \qquad (4-17)$$

for $n-1$ linearly independent constant integer vectors j_r, do all orbits close. This is easy to see: it means that the ratio of any two frequencies is a rational number. A special case is that all frequencies are integer multiples of some one frequency ν_0 (which need not belong to the set $\nu_1, \nu_2, \cdots \nu_n$). Then the orbit (4-16) is a Fourier series in time with fundamental frequency ν_0.

Prob. 4.1. Introduce action-angle variables and solve for the frequency $\nu = \partial K(J)/\partial J$ for

a) the linear harmonic oscillator,

b) the 1D pendulum (4-5). You may want to look at Goldstein [18] for hints.

4.2.2 N-Tori

The moral of Eq. (4-16) is that the bounded motion of integrable systems lies on invariant n-tori. For, once we locate an orbit initially in phase space by the $2n$ numbers q_0, p_0, or equivalently, by the $2n$ numbers θ_0, J_0, the further position is determined uniquely by n angles $\theta_i(t)$, (see (4-15)). n-dimensional manifolds where the point is specified by n *angles* are called *n-tori.* For instance, a 1-torus is a circle; a 2-torus is the surface of a doughnut (Fig. 4.4). For $n \geq 3$, the n-torus cannot be realized in our physical 3-space, but is defined similarly. The n actions $J_i = J_{0i}$, which are constant along the orbit, fix the particular n-torus on which the orbit lies: this torus is *invariant.* In the general case ν_i *incommensurable* ((4-17) not true), the generic orbit winds up on the torus endlessly without closing, and is dense there (as seems reasonable, and can be proved). Only in the commensurable case is every orbit periodic, and therefore not dense on the torus. For "dense," see below.

But this just means that the dynamics of any such integrable system is topologically equivalent to the dynamics of n uncoupled 1D pendula! Indeed, if we pass 2-surfaces through the phase space, they cut the flow in distorted copies of the pendulum phase portrait (Fig. 4.3). This explains the fundamental importance of the pendulum in Hamiltonian dynamics.

This long and rather abstract dose of theory is justified by the pay-off. First, integrable systems can never be "chaotic." For the motion is at most quasi-periodic, and this is too regular for the appellation "chaotic." Secondly, there are implications for the foundations of statistical mechanics. If the microscopic dynamics of nature were integrable, the necessary ergodicity assumptions of statistical mechanics could never be valid. For integrable motion is confined to low-dimensional manifolds, and so could never wander ergodically through the whole energy surface, sampling all states on it, as required by statistical mechanics. That is, $n < 2n - 1$ unless $n = 1$. For a preliminary definition we can say that an orbit is *ergodic* on the energy surface if it is *dense* there: given any state $x \in$ energy surface σ_E and any neighborhood $V \subset \sigma_E$ of x *no matter how small*, the orbit passes through V at some time (or times).

At this point it is natural to entertain the notion that nonintegrable microscopic dynamics will save statistical mechanics. *Viz.*, the nice clean integrable Hamiltonians of our textbooks will be perturbed by realistic interactions into nonintegrable ones. The motion will then be generically ergodic, energy sharing among the

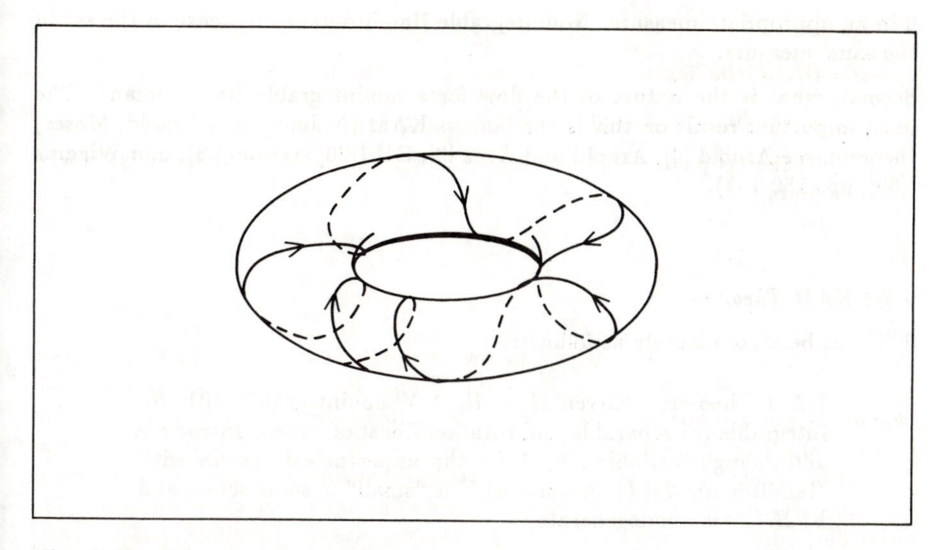

Fig. 4.4 A 2-torus and an orbit on it

various modes (a *mode* is one of the terms in (4-16) characterized by an integer vector j) will take place, equipartition will set in in equilibrium, etc., and all the properties assumed in statistical mechanics will follow. To examine this hopeful view we now turn to nonintegrable systems.

4.3 Nonintegrable Systems

A *nonintegrable* system is simply one which is not integrable. Unfortunately, there is no simple criterion, such that we need only to look at a Hamiltonian to see if it is integrable or nonintegrable. The only way is to prove that the system violates Definition 1, 2, or 3 of section 4.2. But there is no guarantee that we can do this. Consequently, we often work with Hamiltonians which we suspect (from numerical work, say) to be nonintegrable, without being able to prove it.

In this section, we enter an area, called "dynamical systems," on which much research has been done over many years, unlike most work on nonlinear dynamics, which is comparatively new. Some deep and powerful theorems have been proved by great mathematicians. These results are too technical to give in rigorous form in this report. But they are so important that they must be included here, even if in nonrigorous, paraphrased form. My aim, as always in this report, is that the definitions and theorems should make sense to the reader.

First, we might ask: how many integrable Hamiltonians are there among the set of all analytic Hamiltonians? The answer is that they are rare, a set of measure

0 in an appropriate measure. Nonintegrable Hamiltonians are dense in the set in the same measure.

Second, what is the nature of the flow for a nonintegrable Hamiltonian? The most important result on this is the famous KAM (Kolmogorov, Arnold, Moser) theorem, see Arnold [4], Arnold and Avez [3], GH ([20] section 4.8), and Wiggins ([36], pp. 150-153).

4.3.1 KAM Theorem

This can be stated loosely as follows:

> KAM Theorem. Given $H = H_0 + V$ nonintegrable with H_0 integrable (of separable and rotation-libration type). Introduce action-angle variables, θ, J for the unperturbed system with Hamiltonian $H_0(J)$. Assume a) V is "small" in some sense, and b) $H_0(J)$ is nondegenerate,
>
> $$\frac{\partial(\nu_1, \nu_2, \cdots \nu_n)}{\partial(J_1, J_2, \cdots J_n)} \neq 0, \quad \nu_i \equiv \frac{\partial H_0}{\partial J_i},$$
>
> where the left hand side is the Jacobian determinant. Then for a sufficiently irrational set of frequencies $\nu^* \in \Re^n$ there exists an invariant torus $T(\nu^*)$ of the perturbed system H close to the invariant torus $T_0(\nu^*)$ of the unperturbed system H_0. (4 – 18)

Remarks. a) Frequencies ν^* are *sufficiently irrational* if $\nu^* \cdot m$ is bounded away from zero for *all* integer vectors $m \in \Re^n$. More precisely,

$$|\nu^* \cdot m| > \alpha \|m\|^{-\beta} \qquad (4-19)$$

for some fixed α, β and all m, where $\|m\|$ is the length of vector m. Thus not only can no integer linear combination of the ν_i^* vanish, as in (4-17), but these must actually not be too small, in the sense of (4-19). b) Sufficiently near a center of H (section 2.3) the majority of the orbits of H lie on these "preserved" tori, called *KAM tori* (sometimes *nonresonant*, or *irrational, tori*). This is the principal moral of the KAM theorem. On perturbing an integrable system into a nonintegrable one, we do not completely destroy the regular motion — part of the motion remains confined to low (n)- dimensional manifolds and quasi-periodic. c) The KAM theorem says nothing about the fate of tori $T_0(\nu)$ whose frequencies are not sufficiently irrational, that is, about the orbits of H which do not lie on the preserved tori $T(\nu^*)$. Might the motion in this *complement set* be "ergodic" in some sense? See the remarks on statistical mechanics at the end of section 4.2.

4.3.2 Poincaré-Birkhoff Theorem

To investigate the dynamics in this complement set we consider the fate of a *rational torus* $T_0(\nu_R)$ of the unperturbed flow, that is, one whose frequencies are commensurable (4-17), so that any orbit on it is periodic, and closes. We can define a Poincaré map P_0 (section 3.11) by cutting this torus transversely by a $(2n-2)$-dimensional cross section Σ lying in the $(2n-1)$-dimensional relevant energy surface $H_0 = E$. Let Σ cut the torus in the *level curve* Γ. By a simple theorem, every point of Γ is a fixed point of P_0^m for some fixed m; we choose the minimum such positive integer m. Now turn on the perturbation V; the same section Σ will define a Poincaré map P for the perturbed flow near $T_0(\nu_R)$ in the energy surface $H = E$. The question is, what happens to $T_0(\nu_R)$, or equivalently, to Γ?

For a *Hamiltonian map* F we have det $DF(x) = 1$ everywhere, the counterpart of $\nabla \cdot f = 0$ for flows, because it conserves volume in phase space. Hence its fixed points of any order, $F^m(x^*) = x^*$, can be proved to be either centers or saddles in the generic case, cf. the similar discussion for Hamiltonian flows following (4-4). For a map of $\Re^2$ the situation is especially simple. Let $DF^m(x^*)$ have eigenvalues λ_1, λ_2, which thus satisfy $\lambda_1\lambda_2 = 1$. There are only two possibilities: $0 < \lambda_1 < 1 < \lambda_2$, or $\lambda_2 = \lambda_1^*$, $|\lambda_i| = 1$. The fixed point is called a *hyperbolic point* in the first case (saddle point) and an *elliptic point* in the second case (center).

At this point we specialize to two degrees of freedom, $n = 2$. Then we are dealing with a 3D energy surface, a 2D cross section Σ, a 2-torus, a 1D Γ (an ordinary curve), and Poincaré maps P_0 and P taking $\Re^2$ into $\Re^2$. Now we are ready to state the main theorem in a paraphrased form. For rigorous statements, see Arnold and Avez [3], GH ([20], section 4.8), Wiggins ([36], pp. 137-140).

> Poincaré-Birkhoff Theorem. For a sufficiently small perturbation V, the level curve Γ "breaks up" into $2km$ fixed points of P^m for some integer k. These fixed points lie near Γ; km are elliptic and km are hyperbolic. $(4-20)$

This situation is depicted in Fig. 4.5. Γ is often called a *resonant level curve*, and a certain region around Γ containing the elliptic and hyperbolic points is called a *resonance zone.* Also shown are the separatrices joining two adjacent hyperbolic points. If they intersect transversely in a homoclinic point as shown, this leads to the infinitely complicated set of intersections known as a homoclinic tangle (or *stochastic layer* in the physics literature), which was discussed in section 3.6 for a single fixed point. Incidentally, the integer k is not predicted by the P-B theorem. For the situation $km = 3$ of Fig. 4.5, $k = 3$, $m = 1$ or $k = 1$, $m = 3$. The regions around the elliptic points bounded by the separatrices are called *islands.* Figure 4.5 shows an "island chain" of three islands in the case $m = 3$ or a set of three "separate" islands in the case $m = 1$.

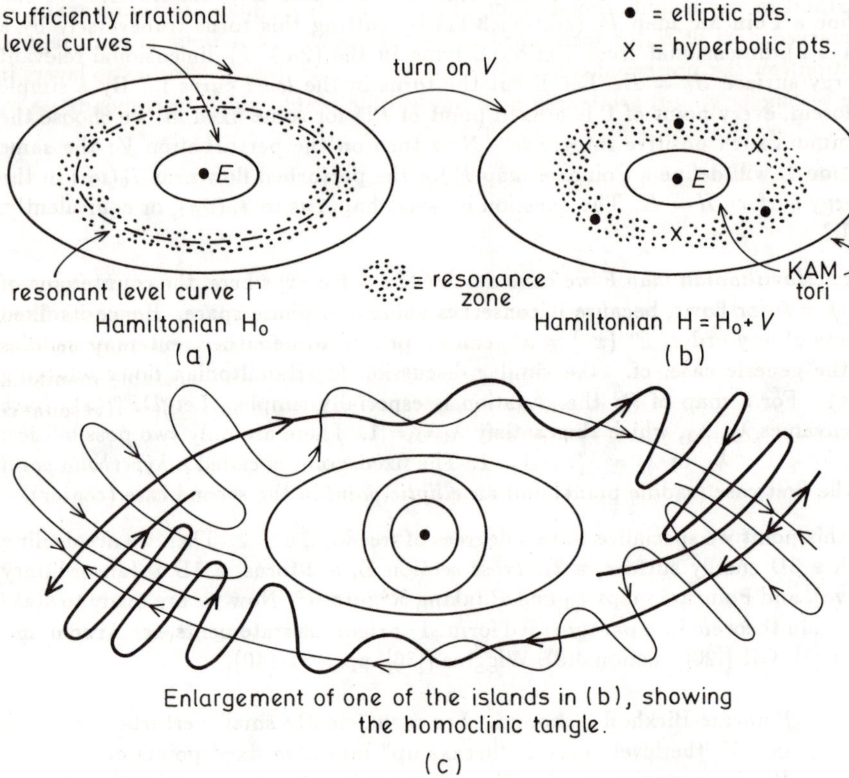

Fig. 4.5 Illustrating the Poincaré-Birkhoff Theorem

Take any elliptic point E of H. In any neighborhood of E there is a resonant torus $T_0(\nu_R)$ of H_0 (Arnold and Avez [3]). We have indicated such an elliptic point E with a nearby resonant level curve Γ on the left side of Fig. 4.5. Thus near any of the three "new" elliptic points shown in Fig. 4.5, call it E', there is a resonant torus $T_0(\nu_R')$, which breaks up as the P-B theorem dictates, producing $k'm'$ new elliptic points $\{E''\}$ (and $k'm'$ new hyperbolic points). Figure 4.5 can represent this breakup by the substitutions $E \to E'$ and $\Gamma \to \Gamma' \equiv$ resonant level curve associated with $T_0(\nu_R')$. But now the P-B theorem can be applied to resonant tori which lie near the $\{E''\}$, producing a breakup of those resonance zones. And so on *ad infinitum*. Thus there is infinitely nested, self-similar structure within any island.

Further, in any neighborhood of an elliptic point of H there are nonresonant, or preserved, tori of H, as mentioned in section 4.3.1. Thus there is both regular motion (on the preserved tori) and "chaotic" motion (in the stochastic layers in the resonance zones) arbitrarily near any elliptic point, which corresponds, as we remember, to a stable periodic orbit of H.

For the generalization to $n > 2$, see the cited works.

4.3.3 Resonance Overlap

The above discussion of the consequences of the P-B theorem (which Poincaré himself found mind-boggling) does not exhaust the complications, the possibilities of even more "chaos." If V is large enough, the resonance zones may *overlap*, defined precisely as the transverse intersection of stable and unstable manifolds $W^s(x_1)$ and $W^u(x_2)$ for x_1 and x_2 hyperbolic points from two *different* resonance zones. This is illustrated in Fig. 4.6.

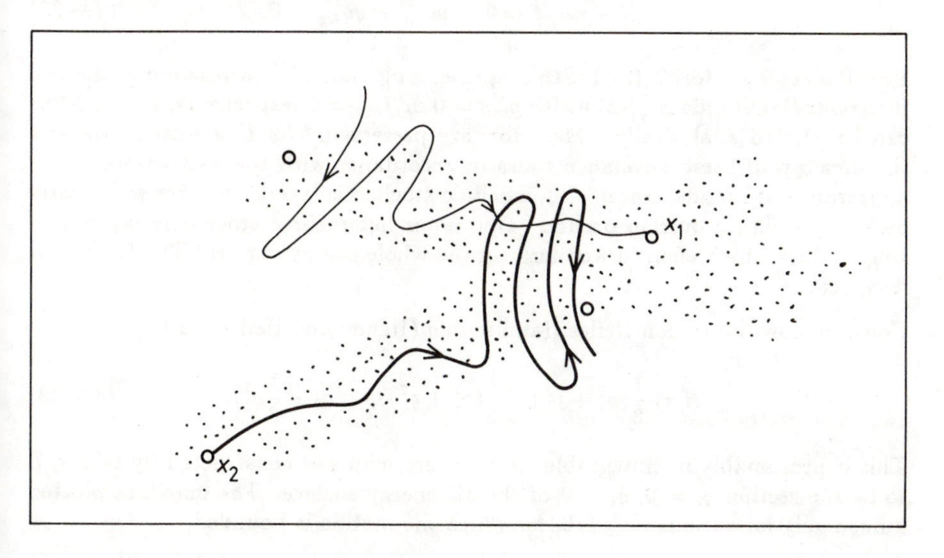

Fig. 4.6 Resonance overlap

Then the inescapable argument of section 3.6 implies infinitely many transverse intersections, thus transverse *heteroclinic* orbits, and *heteroclinic* tangles.

There are theories which offer quantitative methods for handling this P-B breakup of resonance zones, in particular, Melnikov's method, (see GH [20], sections 4.5 and 4.6 and Wiggins [36], section 4.5). For further developments, especially resonance overlap, see Chirikov [13].

4.4 Examples of Nonintegrable Systems

Consider first the "double resonance" Hamiltonian

$$H = H_0(J_1, J_2) + f_{mn}\cos(m\theta_1 - n\theta_2) + g_{mn}\cos(p\theta_1 - q\theta_2) \qquad (4-21)$$

in action angle variables J_i, θ_i, $i = 1, 2$; m, n, p, q are positive integers. For f_{mn} and $g_{mn} \neq 0$, H is presumably nonintegrable, and if they are small enough, the KAM and Poincaré-Birkhoff theorems can be applied. Construction of a Poincaré map P and explicit numerical integration reveal the following features (Walker and Ford [35]).

For low energies E the phase portrait shows the general appearance of the superposition of two resonance zones corresponding to frequencies close to the commensurable ones $\nu_R^{(1)}$ and $\nu_R^{(2)}$ satisfying

$$m\nu_{R1}^{(1)} - n\nu_{R2}^{(1)} = 0, \quad p\nu_{R1}^{(2)} - q\nu_{R2}^{(2)} = 0. \qquad (4-22)$$

Here the resonant tori $T_0(\nu_R)$ with frequencies $\nu_R^{(1)}$ and $\nu_R^{(2)}$ correspond to the two *integrable* Hamiltonians (4-21) with $g_{mn} = 0$ or $f_{mn} = 0$ respectively, whose 2-tori can be plotted analytically. Most tori are preserved. As E is raised, one sees the breakup of these resonance zones in accordance with the P-B theorem, the appearance of islands, traces of homoclinic stochastic layers, etc. For sufficiently high E resonance overlap occurs. One notes heteroclinic stochastic layers and single orbits which seem to wander over the whole energy surface. The KAM tori disappear.

Consider now the Hénon-Heiles Hamiltonian (Hénon and Heiles [22])

$$H = \frac{1}{2}(p_1^2 + p_2^2) + \frac{1}{2}(q_1^2 + q_2^2) + q_1^2 q_2 - \frac{1}{3}q_2^3. \qquad (4-23)$$

This is presumably nonintegrable. A Poincaré map was constructed by taking Σ to be the section $q_1 = 0$, $\dot{q}_1 \geq 0$ of the 3D energy surface. The map was plotted numerically for various $E \leq 1/6$, for which the motion is bounded.

In general one sees preserved tori and broken-up resonance zones. In particular, a) $E = 1/12$. Only 2-tori are seen. b) $E = 0.106$. One sees several large islands and two smaller 8-island chains. There are signs of a stochastic layer near separatrices. c) $E = 0.125$. Some preserved tori. A random splatter of points from a single orbit, possibly ergodic. d) $E = 1/6$. No visible 2-tori. Irregular, possibly ergodic orbits. See Fig. 4.7.

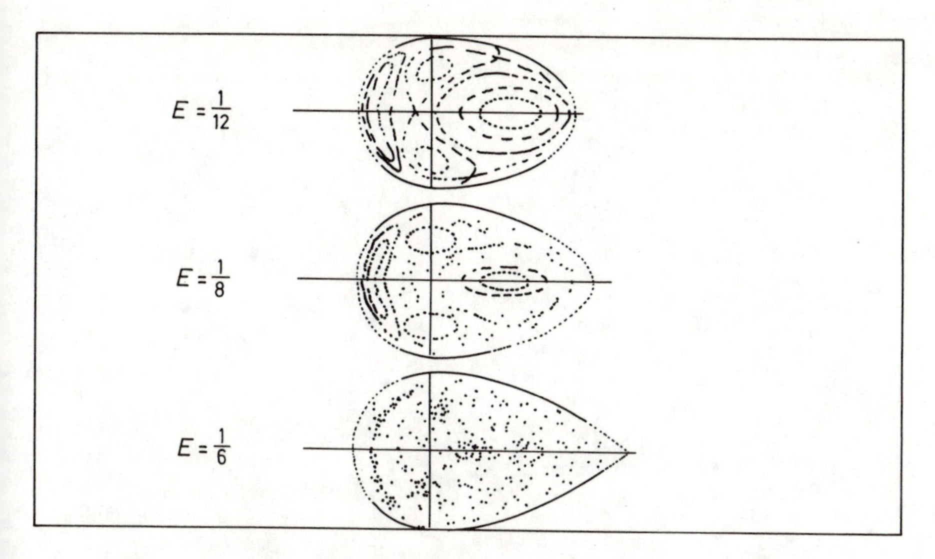

Fig. 4.7 Poincaré map orbits for the Hénon-Heiles system

5. MEASURES OF CHAOS

5.1 Introduction

"Chaos" was already introduced in an informal way in Chap.1, where we saw that although there is no consensus on a precise definition yet, three or four criteria are usually accepted in the physics literature as "signatures of chaos." The core idea of "chaos," however, as most would agree, is sensitive dependence, often abbreviated SD hereafter. We will concentrate on SD in this chapter, and treat the other signatures much more briefly.

The meaning of sensitive dependence (on initial conditions, that is, initial point x_0 in phase space, understood hereafter) is the *unpredictability* of the orbits, even though the dynamics is of course a deterministic system. Say we have a system which does possess SD (not all of them do!) and we wish to compute the orbit $\phi_t(x_0)$ based at x_0. No matter how closely we *approximate* x_0, it is in principle possible (for certain x_0) to make a fixed gross error in the orbit after some time t: the calculated orbit bears *no resemblance* to the true orbit $\phi_t(x_0)$ after this time. The term "unpredictable" seems justified for this property. The necessity of approximating x_0 is forced upon us by inevitable small errors, by the finite resolution of our computers, etc. This situation will not be cured by an increase of 10^3 or 10^6 or ... of our supercomputers of the future. Also, it has nothing to do with noise, errors which arise from perturbations originating outside the system (although noise will compound the error) – it is an intrinsic property of the deterministic system.

There are two versions of SD current today – the mathematician's and the physicist's, which, strangely enough, seem to coexist without being aware of each other. The mathematician's definition focuses on this basic idea of unpredictability, the minimal, irreducible idea, such that anything weaker would be admitted as regular, predictable motion by everybody. It is something very like the familiar idea of the continuity or discontinuity of functions at a point generalized to apply to whole orbits. The mathematician's definition does not concern itself with how fast orbits separate, or how long it takes to make this "fixed, gross error." The physicist's definition on the other hand requires fast (exponential) separation of orbits in a certain sense; we shall call it exponential SD This is partly because there is an easy way (in principle) to check this fast SD, namely by the calculation of Lyapunov exponents. But the main reason, I suspect, is the widely held belief that exponential SD is necessary to ensure "chaos," since the mathematician's SD, call it minimal SD, is known to include cases of "trivial SD," where the behavior is clearly not "chaotic," for example, the free particle. However, this belief is mistaken, as we shall see. Minimal SD occurs on certain sets in phase space; it is natural to limit these sets in a certain way. Then minimal SD, so limited, is essentially equivalent to chaos; it has all the properties that we associate with that term. So chaos can finally be *defined* this way. There is no need to mention the rate of divergence of orbits in the definition, and no reason to believe that all chaotic SD is exponential! The distinction between exponential and minimal SD is irrelevant, a false distinction, as far as chaos goes.

One advantage of the minimal, over the exponential, definition of SD is that it is easier to carry out analytic proofs of its presence or absence in some systems, notably QM (quantum mechanics) systems, in the former than in the latter. If minimal SD can be shown to be *absent*, then there is no chaos in that system. *Proofs* that chaos is absent in a very large class of interesting observables in QM will be given or indicated in this section or in appendix B.

A very interesting and apparently open question (July 1990) is whether chaos defined this way implies the other usually accepted "signatures of chaos."

5.2 Sensitive Dependence

Let x be a point of phase space, or *state*, of the system. For flows and maps whose phase space $\subset \Re^N$, a state is the real N-tuple $(x_1, x_2, \cdots x_N)$. But the state need not be so limited. For example, in QM x is the state vector ψ, sometimes written $|\psi\rangle$, a "point" in an ∞-dimensional complex vector space $\mathcal{H}$. The essential thing is that there should be a *distance* $d(x_1, x_2)$ between any two states x_1, x_2, that is, state space can be any metric space $\mathcal{M}$. For $\Re^N$ we could take $d(x_1, x_2) = \|x_1 - x_2\| \equiv \left[\sum_{i=1}^N (x_{1i} - x_{2i})^2\right]^{\frac{1}{2}}$, the *euclidean metric*. For QM state space $\mathcal{H}$, where $x_1 = \psi_1$, $x_2 = \psi_2$

$$d(x_1; x_2) = \|\psi_1 - \psi_2\|, \quad \text{where } \|\phi\|^2 \equiv \langle\phi|\phi\rangle, \tag{5-1}$$

and $\langle\psi|\phi\rangle$ is the scalar product in Hilbert space $\mathcal{H}$. We can continue to call $\phi_t(x_0)$ the orbit in this general context.

The basic (minimal, irreducible) idea of sensitive dependence at state $x \in \mathcal{M}$ is that there exist states x' *arbitrarily close* to x such that the x' eventually separate from x by at least some fixed finite distance ϵ independent of x' under time evolution. As emphasized in section 5.1, this is it! We do not require exponentially fast separation of $x(t)$ and $x'(t)$. Since this concept is so important, let us put it into a precise definition.

Def. The dynamics D with orbit function ϕ_t has ***sensitive dependence on initial conditions*** at state x if there exists an $\epsilon > 0$ such that, for any neighborhood $\mathcal{N}$ of x there exists an $x' \in \mathcal{N}$ and a $t > 0$ such that $d(\phi_t(x'), \phi_t(x)) > \epsilon$. $(5-2)$

The set of all such x is called the ***Julia set*** J of the dynamics D. This concept is very old, as nonlinear dynamics goes, going back to Julia and Fatou, who worked on the iteration of complex analytic maps in France around 1920. It is clear that Def. (5-2) guarantees the fixed error ϵ in the orbit through an $x \in J$, as discussed qualitatively in section 5.1.

There arises a vexing question here. There may be functions F of the state x which are at least as interesting as x itself. For QM, where the state $x = \psi$, it is sufficient to mention expectation values of Hermitian observables, the time correlation of $\psi(t)$ and $\psi(t+\tau)$, etc. We could define SD for the observable F simply by using $\tilde{d}(F(\phi_t(x')), F(\phi_t(x)))$ in (5-2), where $\tilde{d}$ is the distance in the metric space $\widetilde{\mathcal{M}}$ of the function $F : \mathcal{M} \to \widetilde{\mathcal{M}}$. The question arises: does the SD of the dynamical system D depend on the function, or *map*, F in question? One can prove that it *does not* if $F(x)$ is continuous (Ingraham [24]). Assuming continuous maps F always, we restrict to the orbit itself in this report.

But now notice that the so-defined SD does not always imply what we want to mean by "chaos." A simple example: consider a free particle, moving on a straight line for simplicity. The phase space has points $x = (\xi, v)$, $v \equiv \dot{\xi}$, and orbits $\xi(t) = vt + \xi_0$, $v(t) = v = \text{const}$. Take the euclidean norm $d(x', x) \equiv [(\xi' - \xi)^2 + (v' - v)^2]^{\frac{1}{2}}$. Then it is clear that for any initial point $x \equiv (\xi, v)$ and different point $x' \equiv (\xi', v')$ with $v' \neq v$, we have

$$d(\phi_t(x'), \phi_t(x)) \geq |v' - v|t.$$

Hence, no matter how close x' is to x, $d \geq$ any given ϵ for some time t; if $\epsilon = 1$, $d \geq 1$ for $t = 1/|v' - v|$. So the dynamics has SD at every point in phase space; every point belongs to the Julia set. Similarly, many integrable systems show this "trivial SD" on their unbounded orbits.

There is a very natural way to limit the SD (5-2) to ensure that it is equivalent to chaos. Like all good definitions, it eschews technicalities and a list of special cases; it achieves the desired concept at a single inspired stroke (Wiggins [36], hereafter abbreviated as W). First, let us define SD for ***sets*** Λ with an ϵ the same for all $x \in \Lambda$. Second, let us require that the set Λ be invariant and compact. "Invariant" we already know from Chap. 2; it means that Λ is carried into itself by the time-evolution, so that SD is not a transient property. Λ *compact* means that it is *closed* and *bounded*. Making all of this precise, we end up with the definitions

Def. The dynamics D with orbit function ϕ_t ***has sensitive dependence on the invariant compact set*** Λ if there exists an $\epsilon > 0$ such that for any $x \in \Lambda$ and any neighborhood $\mathcal{N}$ of x there exists an $x' \in \mathcal{N}$ and a $t > 0$ such that $d(\phi_t(x'), \phi_t(x)) > \epsilon$. (5 − 3)

There is another, independent feature of "chaotic" motion on a set Λ, namely that the set be mixed up properly.

Def. For a dynamics D with orbit function ϕ_t, V is *topologically transitive* if, for any pair of open sets $U_1, U_2 \subset V$, the image of U_1 under the dynamics intersects U_2 at some time: $\phi_t(U_1) \cap U_2 \neq \emptyset$for some time $t > 0$. (5 − 4)

Following Wiggins [36] and Devaney [15], we require both (5-3) and (5-4) for true chaos.

Def. The dynamics D is *chaotic* on an invariant, compact set Λ if D has sensitive dependence and is topologically transitive on Λ. (5 − 5)

For convenience of expression we can also say that Λ *is chaotic* in the case (5-5). Thus we have finally a precise definition of chaos in any dynamical system (and we can drop the quotation marks in "chaos" when referring to this definition).

Sensitive dependence on sets (5-3) and chaos can be generalized to maps of the orbit or "observables" as noted above. Again one gets the same SD and chaos with respect to the map as with respect to the orbit itself if $F : \mathcal{M} \to \widetilde{\mathcal{M}}$ is continuous.

Requiring SD on compact sets in phase space as in (5-3) for chaos rather than just pure SD (5-2) excludes those cases of "trivial SD" from classification as chaos. For our free particle example above, the orbits are clearly unbounded, so could never be contained in a *compact* set Λ.

5.2.1 Lyapunov Stability

It is interesting to get an expression of regular (*non*-sensitively dependent) motion by negating the *minimal* definition (5-2) of SD. This reads

Def. The dynamics D is *Lyapunov stable* at state x if, given $\epsilon > 0$, there exists a $\delta > 0$ such that if $d(x', x) < \delta$, then $d(\phi_t(x'), \phi_t(x)) < \epsilon$ for all $t \geq 0$. (5 − 6)

Cf. W, page 6, def. 1.1.1. In the old work on iteration of complex analytic maps $F(z)$, $z \in \mathbf{C}$, mentioned before, such a map F was said to be *normal* at z. Def. (5-6) is the nearest thing to continuity that we can define for whole orbits. Comparing (5-6) with the previous definition of the stability of a fixed point in section 2.3, we see that the latter is none other than Lyapunov stability specialized to the degenerate orbit $x(t) = x^*$ if we define neighborhoods via the metric d in the usual way. So (5-6) is now seen as the logical extension of (2-13a) to whole orbits.

5.2.2 Maximal Lyapunov Exponent

The Lyapunov exponents $\chi_1, \chi_2, \cdots \chi_N$ for a map or a flow in $\Re^N$ can be defined, but this involves some tricky technicalities which don't arise if $N = 1$ (Def. (3-11) for maps). See GH [20]. So we shall skip that in this report and instead define the *maximal Lyapunov exponent* χ for our general phase space $\mathcal{M} \equiv$ any metric space. Let the dynamics have the orbit function ϕ_t.

$$\text{Def.} \quad \chi(x) \equiv \lim_{t\to\infty} \sup_{x' \in \mathcal{N}(x)} \frac{1}{t} \ln \| D\phi_t(x)(x' - x) \| \qquad (5-7)$$

if this limit exists for all sufficiently small neighborhoods $\mathcal{N}(x)$. $D\phi_t(x)$ is the Jacobian matrix of ϕ_t at x for $\Re^N$ or its generalization for more general phase spaces $\mathcal{M}$ such as Hilbert space $\mathcal{H}$ for QM. In (5-7) the Jacobian matrix acts on the vector $x' - x$, and we take the length $\| \cdots \|$ of the resulting vector. *Sup* means least upper bound, here over points x' in sufficiently small neighborhoods of x. Thus we are varying the directions of the small vectors on which the Jacobian matrix acts and taking the maximal vector length so found. A numerical algorithm for computing χ is given by Benettin *et al.* [7].

A little reflection reveals that $\chi(x)$ measures the maximal ratio of exponential divergence of the orbit based at x from other infinitesimally near orbits. $\chi(x) > 0$ for an orbit will be defined as *exponential SD*. Now what we expect intuitively is $\chi(x) \leq 0$ if the orbit through x is Lyapunov stable. I know of no formal proof of this, but it seems very plausible and we shall accept it here. So $\chi(x) > 0$ for some x implies the violation of Lyapunov stability (5-4), so that its negation, minimal SD (5-2), is valid. Exponential SD is therefore just a special case of minimal SD

The motion in homoclinic and heteroclinic tangles has been studied carefully by mathematicians (GH [20], Secs. 5.2, 5.3; W [36], Secs. 4.3, 4.4, 4.8) and proved to possess SD on compact invariant sets Λ, (5-3). These sets Λ, called *Cantor sets*, are really weird and intuition-defying. They are closed, totally disconnected, and *perfect* (every point is a limit point). See the mathematical glossary in GH for these terms. In fact, the motion is topologically equivalent to a certain "shift map" on the space of bi-infinite sequences of N symbols. This *symbolic dynamics* can be proved to be chaotic according to (5-5) (see W section 4.2). However, to my knowledge, it is unknown whether this extremely chaotic motion shows exponential SD

5.2.3 Sensitive Dependence and Chaos in Quantum Mechanics

The question of whether there is "chaos" in QM is one of intense interest at the present moment (1990), hotly disputed, with claims and counterclaims compounding a confused situation. Part of the trouble is, of course, the unsettled nature of the definition of chaos, implicit in my use of quotation marks around this term in most of this report. However, with chaos clearly defined as in (5-5), rigorous proofs that *there is no chaos in many important observables of QM* can be constructed (Ingraham [24]). We treat this topic briefly in this section and give such a proof in appendix B.

Consider the self-correlation $C(t,\tau) \equiv \langle\psi(t)|\psi(t+\tau)\rangle$ or the auto-correlation functional of it

$$C(\tau) \equiv \lim_{T\to\infty} \frac{1}{T}\int_0^T C(t,\tau)\,dt. \qquad (5-8)$$

These have been computed numerically for simple model systems and plotted both in t or τ space and, Fourier-transformed, as power spectra. Decaying behavior in τ and broad band power spectra, "signatures of chaos," have been seen (Milonni *et al.* [30], section 52; Pomeau *et al.* [32]).

However, we can prove that there is no chaos (5-5) in these systems by showing that there is no minimal SD (5-2) in these systems, that is, *that they are Lyapunov stable at every state* $\psi \in \mathcal{H}$. The proof can be done for any QM system – any dimension of $\mathcal{H}$, forced or not, etc. One way would be to show that $C(t,\tau)$ is a continuous function of the orbit $\phi_t(x) \equiv$ the state vector $\psi(t)$ in the Schrödinger Picture, in accordance with the theorem on observables F mentioned above. For the state vector itself almost trivially has no SD We show that it is Lyapunov stable.

Given the orbit $\psi(t)$, let $\psi'(t)$ be an orbit initially near the first one: $\|\psi'(0) - \psi(0)\| < \delta$, where $\|\cdots\|$ is the usual QM norm defined in (5-1). Let $U(t,0)$ be the unitary time-development operator. Then at any time t

$$\|\psi'(t) - \psi(t)\| = \|U(t,0)(\psi'(0) - \psi(0))\| = \|\psi'(0) - \psi(0)\| < \delta \qquad (5-9)$$

because $U(t,0)$ is unitary, does not change the norm of any state, Q.E.D.

But it is probably easier to prove directly that $C(t,\tau)$ is Lyapunov stable. See appendix B. From no SD for $C(t,\tau)$, no SD for $C(\tau)$ in τ follows immediately.

Another map of great importance is the expectation value

$$\langle q\rangle(t) \equiv \langle\psi(t)|\,q\,|\psi(t)\rangle \qquad (5-10)$$

of observable q at time t. For example, the occupation probability of energy eigenstate ϕ_n at time t, $|\langle\phi_n|\psi(t)\rangle|^2$, is the expectation value (5-10) for operator $q \equiv |\phi_n\rangle\langle\phi_n| \equiv P_{\phi_n}$, the projection onto the state ϕ_n. These have been plotted for model systems, both as time series and power spectra, in the literature, and have appeared "chaotic." See, for example, Milonni *et al.* [30] (section 52). However, it can be proved that $\langle q\rangle(t)$ for any QM system and any *bounded* operator shows no SD and hence, no chaos.

Def. Operator q is *bounded* if $\|q\phi\| < M\|\phi\|$, all $\phi \in \mathcal{H}$ (5 − 11)

for some constant $M > 0$. Since this proof is similar to the one for $C(t, \tau)$, we leave it as a problem for the reader.

Prob. 5.1. Prove that $\langle q\rangle(t)$ has no SD for q bounded. (Hint: see appendix B).

It begins to look as if one can prove analytically that there is no chaos in QM! But the situation is not yet clear because of the existence of *unbounded* operators q in QM: (5-11) not true, or equivalently, their spectra unbounded above or below or both. These can't be ignored: a few examples are position and momentum of free particles or oscillators, energies of n-electron atoms, orbital angular momentum, etc. The role that unbounded operators play in possible chaos in QM is an open question (July 1990).

5.3 Discrete Fourier Transform and Power Spectrum

Say we have a set of numbers x_j, $j = 0, 1, 2, \cdots N-1$, where $N >> 1$. They might be obtained from sampling a flow at equally spaced times $j\Delta t$ over a time $T \equiv N\Delta t$, or from a map dynamics $x_j = F^j(x_0)$ over a large number N of iterations. Define the *discrete Fourier transform* of this set by

$$\hat{x}_k \equiv \frac{1}{\sqrt{N}} \sum_{j=0}^{N-1} x_j e^{-2\pi i j k/N}, \quad k = 0, 1, 2, \cdots N-1. \qquad (5-12)$$

$\hat{x}_k$ corresponds to frequency $\nu_k \equiv k/T$. The inverse is

$$x_j = \frac{1}{\sqrt{N}} \sum_{k=0}^{N-1} \hat{x}_k e^{2\pi i k j/N}, \quad j = 0, 1, 2, \cdots N-1, \qquad (5-13)$$

obtained via the "orthogonality relation"

$$\sum_{n=0}^{N-1} e^{2\pi i n(k-j)} = N\delta_{kj}, \quad 0 \le k, j \le N-1. \qquad (5-14)$$

Eq. (5-14) is nothing but the statement that the Nth roots of unity form a closed regular polygon in the complex plane. Then the *discrete power spectrum* P_D is defined

$$P_D(\nu_k) \equiv |\hat{x}_k|^2, \quad k = 0, 1, 2, \cdots N-1. \qquad (5-15)$$

If we want to study a slice of a dynamics (typically in the asymptotic regime) by the discrete Fourier transform, we choose T and Δt such that the important system frequencies are contained in the interval $(1/T, (N-1)/T) \approx (1/T, 1/\Delta t)$.

Incidentally, the fast Fourier transform, or FFT (see Bergé *et al.* [9], section III.4), greatly shortens the numerical evaluation of (5-12).

Several features that distinguish the discrete Fourier transform and its power spectrum from the usual (continuous) Fourier transform and its power spectrum should be noted.

a. If the set $\{x_j\}$ is a slice of a stable periodic orbit of period $\tau << T$, then the discrete power spectrum $P_D(\nu_k)$ has peaks at all harmonics of the fundamental frequency $1/\tau$ less than $\nu_{\max} \approx 1/\Delta t$. For example, for a stable n-cycle of a map, P_D has peaks at the harmonics $m/n, 0 \leq m \leq n-1$, of the fundamental frequency $1/n$. This is a direct reflection of the fact that the Fourier *series* of a process restricted to time interval $(0, \tau)$ or of a process periodic with period τ has nonzero Fourier amplitudes in general at all harmonics of the fundamental frequency $1/\tau$.

b. The peaks in $P_D(\nu_k)$ at the dominant frequencies are finitely high and broad and are asymmetrical. The power spectrum does not vanish between these peaks. It is symmetric around $\nu_{\max}/2$.

c. Contrast the behavior of the *discrete* power spectrum P_D and of the usual *continuous* power spectrum P developed for stationary ergodic random processes (Reif [33], Secs. 15.13 - 15.15). For the quasi-periodic motion (4-16), $P(\nu)$ is a sum of delta functions at the frequencies $\pm\nu_i$, $i = 1, 2, \cdots n$. These are infinitely high and infinitely narrow peaks, and $P(\nu) = 0$ elsewhere. $P_D(\nu_k)$, on the other hand, is peaked not only at the ν_i but also at all their harmonics in the sense of remark (a). These peaks are finitely high and broad, in accordance with remark (b). Call these *spikes.*

The discrete power spectrum is a valuable tool in nonlinear dynamics. Regular motion shows up as spikes. A *broad band power spectrum* $P_D(\nu_k)$, a more-or-less continuous background with a few spikes over it, is accepted as a "signature of chaos."

Further, a *discrete self-correlation*

$$C_D(m) \equiv \frac{1}{N} \sum_{j=0}^{N-1} x_j x_{j+m} \tag{5-16}$$

can be defined. $C_D(m)$ decaying in m corresponds to a broad band power spectrum.

Prob. 5.2 Compute the discrete power spectrum for the logistic map (3-9) and display it on a graph for the values of μ given in Prob. 3.1. Note the different behavior for $\mu < \mu_\infty$ and $\mu > \mu_\infty$, $\mu_\infty \approx 3.5699\cdots$.

5.4 Algorithmic Complexity, Kolmogorov Entropy, Etc.

There are other measures of "chaos," which we shall treat much more briefly than the preceding ones in this section. The interested reader can pursue these topics, lightly touched on here, by going to the cited literature.

5.4.1 Algorithmic Complexity

This is a measure of the *randomness* of the orbits of a dynamics. See Alekseev and Yacobson [1] for details and proofs. Some workers, especially Joseph Ford and his collaborators, equate randomness with chaos. An informal sketch follows.

A *finite word* s_N is a string of N letters from some alphabet $\mathcal{L}$, for example $\mathcal{L} = \{1, 2, \cdots m\}$. A special case is $\mathcal{L} = \{0, 1\}$, the binary alphabet. Then

Def. The *algorithmic complexity* $K(s_N)$ of the finite word s_N is the *length* (number of symbols) of the shortest algorithm which will print the word. (5 – 17)

This definition depends (apparently) on the Turing machine, the algorithm, but can be made machine-independent. *Algorithms* are also called "programs," "recursive relations," or "rules." We have $1 \leq K(s_N) \leq N$ in any case. If, for large N, $K(s_N)$ is not appreciably shorter than N itself: $K(s_N) = O(N)$ as $N \to \infty$, the word is called *random*.

Define the *algorithmic complexity* (AC for short hereafter) of an infinite word s_∞ as

$$\mathcal{K}(s_\infty) \equiv \lim_{N\to\infty} K(s_N)/N, \tag{5 – 18}$$

where s_N is the first N letters of s_∞. Then if $\mathcal{K}(s_\infty) > 0$, the infinite word is called *random*.

Next, consider a map dynamics $F : X \to X$, where the phase space X may be quite general. X is partitioned into a finite number of disjoint sets $\{E_1, E_2, \cdots E_p\} \equiv \mathcal{E}$ (a coarse graining of X). Then the actual orbit $F^n(x)$, $n \geq 0$, is replaced by its *itinerary*: the infinite word $\omega_1, \omega_2, \omega_3, \cdots$ according as the nth point falls into set $E_{\omega_n} : F^n(x) \in E_{\omega_n}$. We then define the AC of this infinite word as in (5-17)! Since the itinerary depends on the initial point x, the partition $\mathcal{E}$, and the map F, denote this AC by $\mathcal{K}(x, F|\mathcal{E})$.

$$\mathcal{K}(x, F|\mathcal{E}) \equiv \lim_{N\to\infty} K(\widetilde{\omega}_N)/N, \tag{5 – 19}$$

where $\widetilde{\omega}_N$ is the finite m-ary sequence $\omega_1, \omega_2, \cdots \omega_N$. Now let $\mathcal{K}(x, F)$ be the least upper bound (or sup) of (5-18) over *all* partitions $\mathcal{E}$. Then we say that the orbit based at x is *random* if $\mathcal{K}(x, F) > 0$.

We can see the motivating idea behind this formalism. If the orbit is really "random" in the intuitive sense, there is no short algorithm available to print it out. We must simply numerically calculate the whole orbit. This makes $\mathcal{K}(x, F) > 0$.

5.4.2 Kolmogorov and Other Entropies

This is an idea closely related to algorithmic complexity, whose aim is to quantify randomness in a dynamics. Again one works with maps F on very general spaces. The *entropy* $h(F)$ measures the uncertainty in predicting the $(N+1)$th point of an itinerary, given the first N points, for $N >> 1$. Again one uses a partition $\mathcal{E}$ of phase space, and replaces an orbit by the labels of the coarse-graining sets that it visits. *Kolmogorov* (or *metric*) entropy uses measure theory and a generic orbit. The *topological entropy* is defined slightly differently, following individual orbits of F. See Alekseev and Yacobson [1] and Young [37]. The upshot is that the dynamics is called random or unpredictable if the entropy $h(F) > 0$.

To my knowledge, the precise interrelationships of sensitive dependence, algorithmic complexity, and the various entropies have never been clarified. There are scattered cross-connections like the Pesin formula, (see Young [37], p. 604). The ideas of unpredictability, randomness, and chaos certainly overlap; they cannot be independent. For instance, one could ask: does a dynamics which shows SD in the minimal sense (5-2) necessarily have positive algorithmic complexity and entropy? My conjecture is "Yes."

6. RENORMALIZATION GROUP

6.1 Introduction

There are phenomena in mathematics, physics, and other sciences which occur repetitively on all length scales. The geometry of such invariant limit sets is called *self-similar* (precisely, the part is topologically equivalent to the whole under a scale change $x \mapsto \alpha x$, $\alpha = \text{const.}$). We already saw an example of this in the period-doubling process in the logistic map (section 3.10.2). An all-important example is the phenomenon of phase transitions in condensed matter. At the critical point the interactions are self-similar under arbitrary changes of scale ("coarse-grainings") because the correlation length is strictly infinite at the critical point. The clearest explanation of this in nontechnical language that I know is Bruce and Wallace [10]. This directly leads to an algorithm for computing the famous *critical exponents* of thermodynamics, for example, the exponent α in the heat capacity C near the critical temperature T_c,

$$C \propto |T - T_c|^{-\alpha} \quad \text{for } |T - T_c| \text{ small,} \tag{6 - 1}$$

which has a typical value $\alpha \approx 0.125$.

This algorithm, known as the *renormalization group* (RG), exploits the self-similarity by defining a transformation, or map, R of coupling constants. If R is iterated, the system is taken into the asymptotic, noncritical regime of weak coupling constants, where known analytic results or perturbation theory give reliable results. Iterating backward, we recover desired thermodynamic variables such as a free energy in the realistic coupling range. The RG can thus in principle deliver exact results inaccessible to perturbation theory. The system at the critical point x_c itself is not changed under the iteration because of the self-similarity there: $R(x_c) = x_c$, that is, x_c is a *fixed point* of the RG coupling constant map R. This ties phase transitions to nonlinear dynamics, its concepts and its techniques. In particular, linear stability analysis at $x^* = x_c$ gives the critical exponents. For a simple treatment of the mathematics, see Maris and Kadanoff [29] or Chandler [12] (Secs. 5.6, 5.7); for the more advanced and complete theory see Plischke and Bergersen [31] (Chap. 6).

6.2 RG for Ising Lattices

Ising lattices are first of all *lattices*: regular nD repeating patterns of discrete points. Second, there are *spins* s_i, $i = 1, 2, 3, \cdots N$, at the N lattice sites, each having only two values "up" or "down": $s_i = \pm 1$, all i. A particular configuration $\{s_1, s_2, \cdots s_N\} \equiv \nu$, where each $s_i = +1$ or -1, is a (microscopic) *state* of the system. Thus there are 2^N states. The energy is due to interactions between *nearest neighbor* spins in the lattice; the spins can also interact with an external magnetic field H via their magnetic moments μ. Thus we can write for the energy E_ν of the state ν

$$E_\nu = -J \sum{}' s_i s_j - \mu H \sum_i s_i \quad , \qquad (6-2)$$

where the prime on $\sum$ restricts the sum to nearest neighbor *pairs*. $J > 0$ corresponds to a model of a ferromagnet, but Ising lattices model very many systems, even "lattice gases," under a slight reinterpretation.

The basic problem in equilibrium statistical mechanics is to evaluate the *partition function* Z,

$$Z \equiv \sum_\nu e^{-\beta E_\nu} \ , \ \beta \equiv 1/kT \ , \qquad (6-3)$$

where $T \equiv$ absolute temperature and $k \equiv$ Boltzmann's constant. The sum is over *all* states ν. (The grand partition function will not enter here.) From Z the entire thermodynamics follows!

For dimension $n = 1$ (6-2) can be written

$$E_\nu = -J \sum_{i=1}^{N} s_i s_{i+1} - \mu H \sum_{i=1}^{N} s_i \qquad (s_{N+1} \equiv s_1), \qquad (6-4)$$

where for convenience the 1D lattice ("chain") was closed by imposing periodic boundary conditions. Then

$$Z(K, h, N) \equiv \sum_{\{s_i\}} \exp(K s_i s_{i+1} + h s_i), \qquad (6-5)$$

where $K \equiv \beta J$, $h \equiv \beta \mu H$, and the sum is over all 2^N states $\{s_i\}$. The sum can actually be done in closed form (!) (see Plischke and Bergersen [31], section 3.F).

> **Prob. 6.1** Evaluate (6-5). This is mainly to get you thinking about such large sums and to see how nontrivial even the simple-looking sum in (6-5) is! For the transfer matrix method, see Plischke and Bergersen, or Chandler [12] (Ex. 5.21).

For $h = 0$ the answer is especially simple,

$$Z(K, N) \equiv (2 \cosh K)^N \qquad (6-6)$$

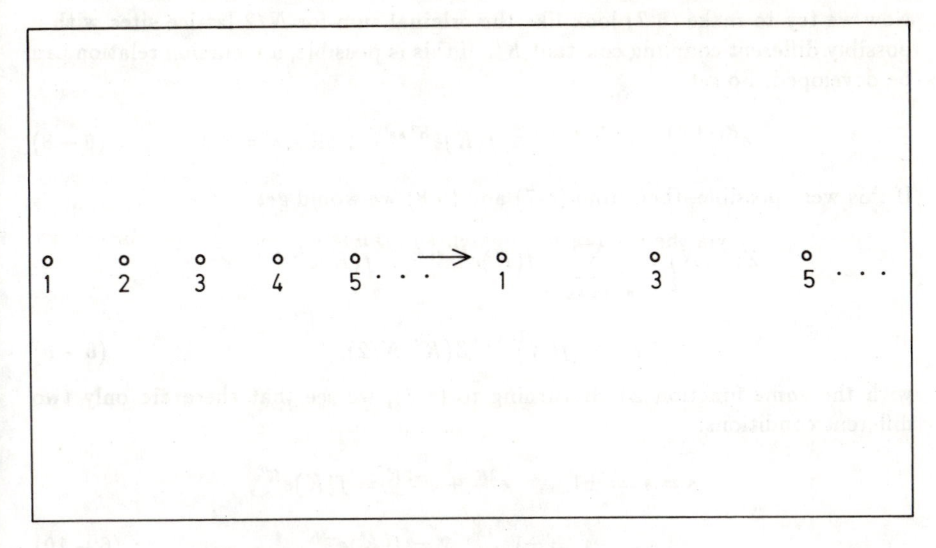

Fig. 6.1 "Thinned out" Ising chain (see text)

(A clarification: this is the answer in the so-called *thermodynamic limit* $N \to \infty$, that is, the leading term.) Then it turns out that the $1D$ Ising lattice has no (proper) phase transition: the order parameter $m \equiv \langle s_i \rangle$, any i, the mean value of the spin for $h = 0$, stays zero for all $T > 0$.

6.2.1 1D Ising Lattice by the RG

Now we wish to derive (6-6) by a recursive scheme, the RG, and verify that there is no phase transition. We follow Maris and Kadanoff [29] and Chandler [12].

Write out the sum (6-5) with $h = 0$ and group the factors in a certain way:

$$Z(K, N) = \sum_{s_1, s_2, s_3, \cdots} \exp[K(s_1 s_2 + s_2 s_3 + s_3 s_4 + \cdots)]$$

$$= \sum_{s_1, s_2, s_3, \cdots} \exp K(s_1 s_2 + s_2 s_3) \exp K(s_3 s_4 + s_4 s_5) \cdots .$$

Sum first over even-numbered spins 2, 4, 6, $\cdots$.

$$Z(K, N) = \sum_{s_1, s_3, s_5, \cdots} \{\exp K(s_1 + s_3) + \exp(-K)(s_1 + s_3)\} \times$$
$$\{\exp K(s_3 + s_5) + \exp(-K)(s_3 + s_5)\} \times \cdots . \qquad (6-7)$$

Every other degree of freedom has been removed; the spins have been "thinned out" (see Fig. 6.1).

Now we try to make (6-7) look like the original sum for $N/2$ lattice sites with a possibly different coupling constant K'. If this is possible, a recursion relation can be developed. So set

$$e^{K(s+s')} + e^{-K(s+s')} \equiv f(K)e^{K'ss'} \quad , \text{ all } s, s' = \pm 1. \tag{6-8}$$

If this were possible, then from (6-7) and (6-8) we would get

$$Z(K,N) = \sum_{s_1,s_3,s_5,\cdots} f(K)e^{K's_1s_3} \times f(K)e^{K's_3s_5} \times \cdots$$

$$\equiv [f(K)]^{N/2} Z(K', N/2), \tag{6-9}$$

with the *same* function Z! Returning to (6-8), we see that there are only two different conditions:

$$s = s' = \pm 1 \quad , \quad e^{2K} + e^{-2K} = f(K)e^{K'},$$

$$s = -s' = \pm 1 \quad , \quad 2 = f(K)e^{-K'}. \tag{6-10}$$

These can be solved, and we get

$$K' = \frac{1}{2}\ln\cosh 2K \quad , \quad f(K) = 2(\cosh 2K)^{1/2}. \tag{$\mathcal{A}$}$$

Now by extensivity we can set $\ln Z(K,N) \equiv Ng(K)$. Since the (Helmholtz) free energy $G \equiv -kT\ln Z$ in general, we see that $g = -\beta\, G/N$, apart from the factor $-\beta$ the free energy per spin. We call g "the free energy" from now on. The natural logarithm of (6-9) can then be rewritten $g(K) = (1/2)\ln f(K) + (1/2)g(K')$, or, with $(\mathcal{A})$, as

$$g(K') = 2g(K) - \ln[2(\cosh 2K)^{1/2}]. \tag{$\mathcal{B}$}$$

Equations $(\mathcal{A})$ and $(\mathcal{B})$ are called the RG equations of the 1D Ising lattice. By noting $f(K) = 2\exp K'$ from $(\mathcal{A})$, the inverse transformation can be written

$$K = \frac{1}{2}\cosh^{-1}(e^{2K'}) \quad , \tag{$\mathcal{C}$}$$

$$g(K) = \frac{1}{2}g(K') + \frac{1}{2}\ln 2 + \frac{K'}{2} \quad . \tag{$\mathcal{D}$}$$

We note immediately that $K' < K$ always.

The first application we leave as a problem.

Prob. 6.2 Apply $(\mathcal{C})$ and $(\mathcal{D})$ starting with small K' and iterate to get $g(K)$ for "realistic" values of $K \approx 2.7$. Take say $K' = 0.01$, for which you can use $Z(K', N) \approx Z(0, N) = 2^N$ from the exact result (6-6), and thus

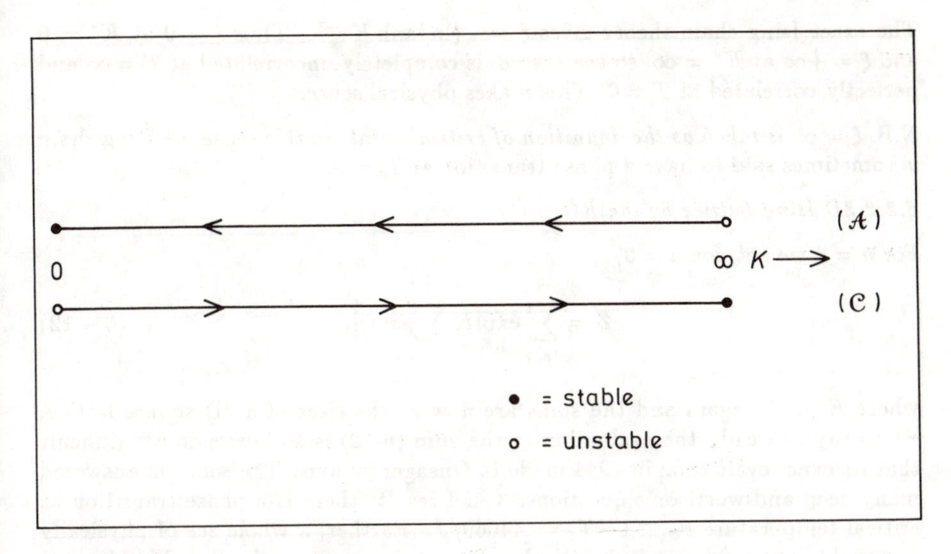

Fig. 6.2 Phase portraits for the 1D maps $\mathcal{A}$ and $\mathcal{C}$, showing stable and unstable fixed points

> $g(K') \approx \ln 2$. Compare $g(K)$ at each step with the exact value derived from (6-6). Notice how fast the iteration converges to the exact values.

On the other hand, we could start with large K, taking $Z(K,N) \approx e^{NK}$ and $g(K) \approx K$ from (6-6), and iterate ($\mathcal{A}$) and ($\mathcal{B}$) to try to calculate g for medium K. But this does not work, the errors grow exponentially!

Second application. Examine the "flow" (used here for a map orbit) of the coupling constant under ($\mathcal{A}$) and ($\mathcal{C}$). See Fig. 6.2.

$K^* = 0, \infty$ are the *only* fixed points of ($\mathcal{A}$) or its inverse ($\mathcal{C}$). Point 0 is stable, point ∞ is unstable under the map ($\mathcal{A}$): $K' = R(K) \equiv (1/2) \ln \cosh 2K$, as you can check. What this means physically is as follows: on thinning out the spins (going to a larger *length scale* $\equiv$ lattice spacing), we go to a system with a weaker coupling constant $K' < K$. This is the meaning of the flow in Fig. 6.2. Only the points $K = 0$ and ∞ are independent of this change of scale since $K^* = 0$ ($T = \infty$) is a completely disordered, $K^* = \infty$ ($T = 0$), a completely ordered, regime. A proper phase transformation would correspond to a finite K^* ($\neq 0$ or ∞) also independent of a change of scale, thus a fixed point K^* of $R \neq 0$ or ∞. The absence of such a finite fixed point in addition to the *trivial fixed points* $K^* = 0, \infty$ of the map R means that the Ising chain shows no phase transition.

One can define a *correlation length* ξ by

$$\langle s_1 s_j \rangle \equiv e^{-j/\xi} \quad , \; j = 1, 2, \cdots N. \tag{6-11}$$

The exact Ising chain theory gives $\xi = -(\ln \tanh K)^{-1}$. Thus $\xi = 0$ at $K^* = 0$ and $\xi = +\infty$ at $K^* = \infty$, so the system is completely uncorrelated at $T = \infty$ and perfectly correlated at $T = 0$. This makes physical sense.

N.B. $\xi = \infty$ *is taken as the definition of critical point.* In this sense the Ising chain is sometimes said to have a phase transition at $T_c = 0$.

6.2.2 2D Ising Lattice by the RG

For $n = 2$ we get, for $h = 0$,

$$Z = \sum_{\{s_i\}} \exp[K \sum{}' s_i s_j], \qquad (6-12)$$

where $K \equiv \beta J$ again and the spins are now at the sites of a 2D square lattice, let us say. Already the evaluation of the sum (6-12) is so horrendously difficult that its exact evaluation in 1944 made L. Onsager famous. The solution answered many deep and worrisome questions. Chief result: there *is* a phase transition at critical temperature $\beta_c \equiv 1/kT_c = .44069/J$. Further, a whole set of physically reasonable thermodynamical critical exponents was obtained. See Plischke and Bergersen [31] (section 5.A) for details.

Now we try to apply the RG method to (6-12). We proceed just as in section 6.2.1, so we can be briefer. Write out (6-12) and group the factors into sets each of which contains just one spin from every other diagonal, spins $\cdots 5, 6, \cdots$ (see Fig. 6.3). Then sum over spins $\cdots 5, 6, \cdots$.

$$Z(K,N) = \sum_{s_1, s_2, s_3, \cdots} \exp[Ks_5(s_1+s_2+s_3+s_4)] \exp[Ks_6(s_2+s_3+s_7+s_8)] \cdots$$

$$= \sum_{\text{remaining } s_i} \{\exp[K(s_1+s_2+s_3+s_4)] + \exp[(-K)(s_1+s_2+s_3+s_4)]\} \times$$

$$\{\exp[K(s_2+s_3+s_7+s_8)] + \exp[(-K)(s_2+s_3+s_7+s_8)]\} \times \cdots. \qquad (6-13)$$

But now we cannot set the first factor $\{\cdots\}$ in (6-13) equal to $f(K)\exp[K'(s_1 s_2 + s_2 s_3 + s_3 s_4 + s_4 s_1)]$ (that is, nearest neighbors in the thinned out lattice) for all $s_i = \pm 1$, $i = 1, 2, 3, 4$, since this gives four independent equations, and we have only $f(K)$ and K' to choose. So we impose

$$\exp[K(s_1+s_2+s_3+s_4)] + \exp[(-K)(s_1+s_2+s_3+s_4)]$$

$$= f(K)\exp[\frac{1}{2}K_1 \underbrace{(s_1 s_2 + s_2 s_3 + s_3 s_4 + s_4 s_1)}_{nn} + K_2 \underbrace{(s_1 s_3 + s_2 s_4)}_{nnn} + K_3 \underbrace{s_1 s_2 s_3 s_4}_{square}], \qquad (6-14)$$

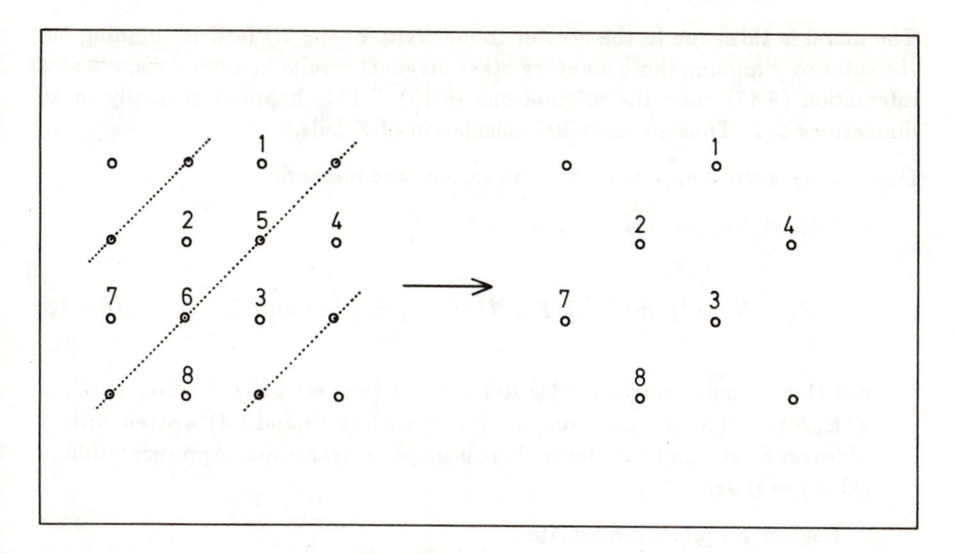

Fig. 6.3 "Thinned out" 2D square Ising lattice

where we have labeled the terms *nn* for nearest neighbors, *nnn* for next nearest neighbors, and *square* for the four spins around a square in the coarse-grained lattice, see Fig. 6.3. Then the four independent possibilities: all s_i equal, three s_i equal and one unequal, $s_1 = s_2 = -s_3 = -s_4$, and $s_1 = -s_2 = s_3 = -s_4$, $(i = 1,2,3,4)$ give the four equations

$$\begin{aligned} \exp 4K + \exp(-4K) &= f(K)\exp(2K_1 + 2K_2 + K_3), \\ \exp 2K + \exp(-2K) &= f(K)\exp(-K_3), \\ 2 &= f(K)\exp(-2K_2 + K_3), \\ 2 &= f(K)\exp(-2K_1 + 2K_2 + K_3). \end{aligned} \qquad (6-15)$$

The solution is

$$K_1 = \frac{1}{4}\ln\cosh 4K, \quad K_2 = \frac{1}{8}\ln\cosh 4K, \quad K_3 = \frac{1}{8}\ln\cosh 4K - \frac{1}{2}\ln\cosh 2K,$$

$$f(K) = 2(\cosh 2K)^{\frac{1}{2}}(\cosh 4K)^{\frac{1}{8}}. \qquad (6-16)$$

Now substitute (6-14) for each factor $\{\cdots\}$ of (6-13). A factor $f(K)^{N/2}$ comes outside. Notice each *nn* pair appears twice, while each *nnn* pair and set of four spins in a square appears once. Hence

$$Z(K,N) = [f(K)]^{N/2} \sum_{N/2 \text{ spins}} \exp[K_1 \sum{}' s_i s_j + K_2 \sum{}'' s_l s_m + K_3 \sum{}''' s_p s_q s_r s_t]. \qquad (6-17)$$

Sum $\sum'$ is over all *nn* pairs, sum $\sum''$ is over all *nnn* pairs, sum $\sum'''$ is over all sets of four spins in a square in the coarse-grained lattice.

The moral is this: due to the greater connectivity of the 2D lattice, thinning out the spins by summing those on every other diagonal results in a more complicated interaction (6-17) than the original one (6-12). This happens generally in all dimensions > 1. Thus an exact RG calculation of Z fails.

One can try several approximations to salvage the method.

1. Neglect K_2, K_3 entirely. Then

$$Z(K,N) \approx [f(K)]^{N/2} Z(K_1, N/2) \, , \; K_1 = \frac{1}{4} \ln \cosh 4K. \qquad (6-18)$$

But this is equivalent to the 1D RG analysis (just set $2K \equiv \widetilde{K}$, $2K_1 \equiv \widetilde{K}'$, $Z(K,N) \equiv \widetilde{Z}(\widetilde{K},N)$, and compare (6-18) with (6-9) and ($\mathcal{A}$) written with tildes on K, K', and Z.) Hence there is no phase transition. Approximation (1) is too rough.

2. Neglect K_3 and approximate

$$K_1 \sum{}' s_i s_j + K_2 \sum{}'' s_l s_m \approx K'(K_1, K_2) \sum{}' s_i s_j \qquad (6-19)$$

for some $K'(K_1, K_2)$. This gives (6-18) with this K' substituted for K_1. Introduce the free energy $g(K)$ as before; then the ln of this last equation gives

$$g(K') = 2g(K) - \ln\{2[\cosh 2K]^{\frac{1}{2}} [\cosh 4K]^{\frac{1}{8}}\}, \qquad (B)$$

where the last equation of (6-16) was used to substitute for $f(K)$. Now (6-19) cannot, of course, be satisfied for all spin values. We want to estimate K' by satisfying it when all spins are aligned. Since there are N *nn* and *nnn* pairs in a 2D square lattice of $N/2$ sites, we have in the aligned case

$$K_1 \sum{}' s_i s_j = NK_1 \, , \; K_2 \sum{}'' s_l s_m = NK_2 \, ,$$

thus $K_1 + K_2 = K'$ from (6-19). Then from (6-16),

$$K' = \frac{3}{8} \ln \cosh 4K \qquad (A)$$

To summarize: (A) and (B) are our approximate RG equations for the 2D Ising lattice. The inverses can be written down, but we skip them.

Now the map (A) has trivial fixed points at $K^* = 0, \infty$ as before. But it has the nontrivial fixed point $K^* = K_c$,

$$K_c = \frac{3}{8} \ln \cosh 4K_c \Rightarrow K_c \approx 0.50698. \qquad (6-20)$$

But now $K^* = 0$ and ∞ are stable, while K_c is unstable. See the phase portrait of the "flow" in Fig. 6.4.

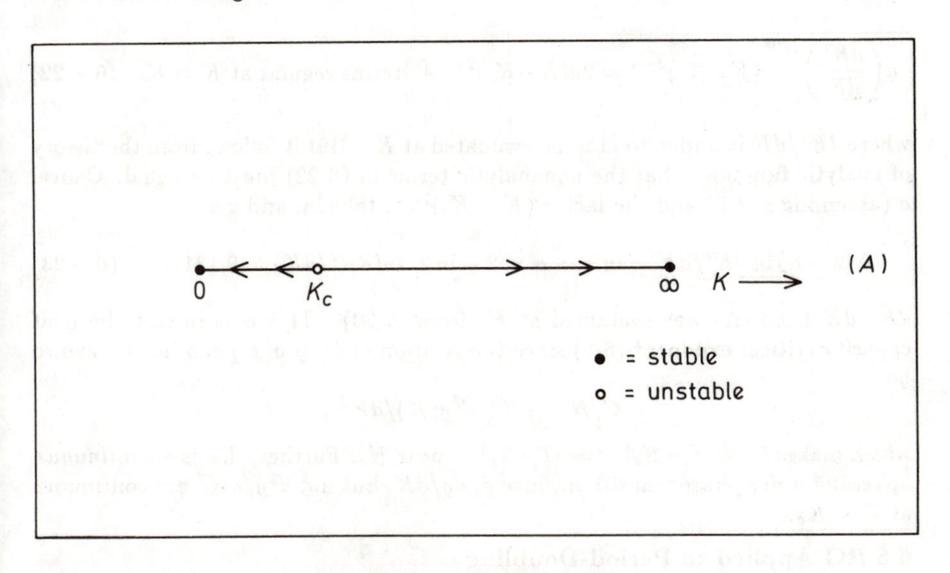

Fig. 6.4 Phase portrait of the 1D map A, showing the trivial stable fixed points and the nontrivial unstable critical fixed point

The physical meaning of this parallels the discussion at the end of section 6.2.1. Going to a larger length scale now drives systems with coupling weaker than K_c to the completely disordered regime $K^* = 0$, but systems with coupling stronger than K_c are driven to the completely ordered regime $K^* = \infty$. But systems with $K = K_c$ exactly are independent of a change of scale, have correlation length $\xi = \infty$, and correspond to a proper phase transition. According to (6-20), this critical temperature T_c is given by

$$J/kT_c \approx 0.50698 \; ; \qquad (6-21)$$

compare the exact value 0.44069 from Onsager's solution.

Finally let us see how critical exponents are predicted on the RG approach. We *assume* that the free energy $g(K)$ has a nonanalytic part which $\approx a(K-K_c)^{2-\alpha}$ near K_c. Then from (B) above and

$$K' - K_c \approx (dK'/dK)\Big|_{K_c} (K - K_c)$$

we get

$$a\left(\frac{dK'}{dK}\right)^{2-\alpha} (K-K_c)^{2-\alpha} = 2a(K-K_c)^{2-\alpha} + \text{ terms regular at } K = K_c. \quad (6-22)$$

where dK'/dK is understood to be evaluated at K_c. But it follows from the theory of analytic functions that the nonanalytic terms in (6-22) must be equal. Cancel a (assuming $a \neq 0$) and the factor $(K-K_c)^{2-\alpha}$, take ln, and get

$$(2-\alpha)\ln dK'/dK = \ln 2 \Rightarrow \alpha = 2 - \ln 2/\ln(dK'/dK) \approx 0.131. \quad (6-23)$$

dK'/dK from (A) was evaluated at K_c from (6-20). This α is in fact the heat capacity critical exponent (6-1) since the relation of C to $g(K)$ can be checked to be

$$C/N = kK^2 \; d^2g(K)/dK^2 \,,$$

which makes $C \sim |K-K_c|^{-\alpha} \sim |T-T_c|^{-\alpha}$ near K_c. Further, this is a *continuous*, or *second order* phase transition, since g, dg/dK, but not d^2g/dK^2 are continuous at $K = K_c$.

6.3 RG Applied to Period-Doubling

We treat the 1D quadratic map with parameter A,

$$F_A \equiv x^2 + Ax \quad , -2 < A < 1\,, \quad (6-29)$$

which maps the interval $[-1, 1-A]$ into itself. This map is topologically equivalent to the logistic map (3-9) by a linear change of variable. There are two fixed points $x^* = 0, -A$. In $-1 < A < 1$ both are stable (remember, "stable" will mean asymptotically stable in this section). At $A = -1$ both go unstable, and the stable fixed points

$$x^*_\pm = -\frac{1}{2}(A+1) \pm \frac{1}{2}\sqrt{(A+1)(A-3)} \quad (6-30)$$

of F^2 are born. These form a stable 2-cycle of F. Consider the orbit based at $x^*_+ + u$, u small. Then

$$x_1 \equiv F(x^*_+ + u) = x^*_- + \Delta x_1,\; \Delta x_1 \equiv (A + 2x^*_+)u + u^2, \quad (6-31)$$

where we simply expanded (6-29) out and used $F(x_+^*) = x_-^*$. Iterating again, we get obviously

$$x_2 \equiv F(x_-^* + \Delta x_1) = x_+^* + \Delta x_2, \qquad \Delta x_2 \equiv (A + 2x_-^*)\Delta x_1 + (\Delta x_1)^2 \ , \quad (6-32)$$

where we used (6-31) and $F(x_-^*) = x_+^*$. Now substitute Δx_1 from (6-31) into Δx_2 from (6-32), and keep only terms of $O(u^2)$. By expanding out, this becomes

$$\Delta x_2 \approx (A + 2x_-^*)(A + 2x_+^*)u + [A + 2x_-^* + (A + 2x_+^*)^2]u^2. \qquad (6-33)$$

This looks something like the original map (6-29); by rescaling

$$x' \equiv \alpha u \ , \ x_2' \equiv \alpha \Delta x_2, \quad \alpha \equiv A + 2x_-^* + (A + 2x_+^*)^2, \qquad (6-34)$$

(6-33) takes the form $x_2' = F_{A'}'(x')$, where

$$F_{A'}'(x') \equiv x'^2 + A'x', \ A' \equiv (A + 2x_-^*)(A + 2x_+^*) = -A^2 + 2A + 4. \qquad (6-35)$$

We used $x_\pm^*$ from (6-30). But $F_{A'}'$ is the *same function* which occurred in the original map (6-29) for a transformed parameter

$$A' = R(A) \equiv -A^2 + 2A + 4 \, . \qquad (6-36)$$

Thereby we have achieved an RG transformation for period-doubling in quadratic maps. This guarantees the self-similarity noted in section 3.10.2, the period-doubling cascade.

An important fine point: the map (6-35) is not the *same* as the map (6-29) since different regions are mapped, cf. (6-34) and $u \equiv x - x_+^*$. That is, F' is not the *same* map as F; rather, they are *topologically equivalent* !

Now we can draw various conclusions. The fixed points $x_\pm^*$ of F_A^2 are the same states as the fixed points $0, -A'$ of $F_{A'}'$. Thus $x_\pm^*$ are stable in the range $-1 < A' < 1$, or $1 - \sqrt{6} < A < -1$ from (6-36). At $A = 1 - \sqrt{6} \approx -1.4495$ they lose stability, and two stable fixed points of F'^2 are born, that is, a stable 4-cycle of F. Thus the RG method can approximate the period-doubling bifurcation values which we called μ_n, $n \geq 3$, for the logistic map in section 3.10.

We can continue this way, defining period-doubling bifurcation values A_n with $A_1 \equiv 1 - \sqrt{6}$. In the limit $n \to \infty$, $A_n \to$ a limit A_∞. Solving $A_\infty = -A_\infty^2 + 2a_\infty + 4$ from (6-36), we get

$$A_\infty = \frac{1}{2}(1 - \sqrt{17}) \approx -1.5616 \, . \qquad (6-37)$$

Prob. 6.3 Find the linear transformation connecting map (6-29) with the logistic map. Thus translate the values A_1 and A_∞ found here into values of μ, and compare with μ_3 and μ_∞ found in section 3.10.

If we assume that the convergence of A_n is asymptotically geometric: $A_n = A_\infty + C\delta^{-n}$ $(\delta > 1)$, we get the *same* δ found before, (3-10). This is no accident; one can prove easily that if we have a new parameter $A = g(\mu)$, g differentiable, then δ_g defined by (3-10) with $\mu \to g(\mu)$ is exactly the same: $\delta_g = \delta$. Further, the scaling parameter α defined in (6-34) should also be asymptotically universal (check by putting $x^*_\pm$ from (6-30) and $A = A_\infty$ into this formula). The "exact" value found for the logistic map is $\alpha \approx -2.5029078\cdots$.

6.4 RG Applied to the Tangent Bifurcation

Tangent bifurcations were described in section 3.10.3. We want now to calculate the time of transit, the time that the orbit point stays in the "neck" in Fig. 3.9, by RG methods.

Iterate the map G_μ and, in keeping with the spirit of the model, keep only terms of $O(x^2)$ and $O(\mu)$.

$$G^2_\mu(x) \equiv \mu + (\mu + x - x^2) - (\mu + x - x^2)^2 \approx 2\mu + x - 2x^2. \qquad (6-38)$$

But

$$\frac{1}{2}G_{4\mu}(2x) = 2\mu + x - 2x^2 \quad ,$$

the same as (6-38), so

$$G^2_\mu(x) \approx \frac{1}{2}G_{4\mu}(2x). \qquad (6-39)$$

This is exactly the statement that the map G^2_μ is topologically equivalent to the map $G_{4\mu}$ in the region $X : x$ and μ small via the homeomorphism $h(x) \equiv 2x$, cf. (3-7).

Hence if we define: $N(\mu)$ steps are necessary for the orbit point of G_μ to pass through the neck, we have

$$N(\mu)/2 = N(4\mu) \quad , \qquad (6-40)$$

since the transit time is only half as long for G^2_μ as for G_μ. This functional equation has the solution

$$N(\mu) = C/\sqrt{-\mu}\,,\ C = \text{const.} \quad (\mu < 0), \qquad (6-41)$$

which is the desired answer.

6.5 RG Applied to Percolation

Percolation theory has evolved considerably since its humble beginnings as the study of the percolation of fluids through porous solids. For some basics, see Plischke and Bergersen [31] (pp. 317, 318). The notions we need are the following: there is a lattice, say 2D, with site occupation probability p. At $p = p_c$, the *critical*

probability, the infinite cluster just forms. At this point the ***configuration***: the lattice with occupied and unoccupied sites marked • and ∘ respectively, should be self-similar, that is, be invariant under an arbitrary change of scale $x \mapsto \alpha x$, (see section 6.1).

We take a triangular lattice with some occupation probability p and coarse-grain it by dividing it into blocks of three sites. These ***block sites*** will be said to be occupied or not by the ***majority rule***: occupied if the number of •'s is three or two, unoccupied if this number is one or zero. See Fig. 6.5.

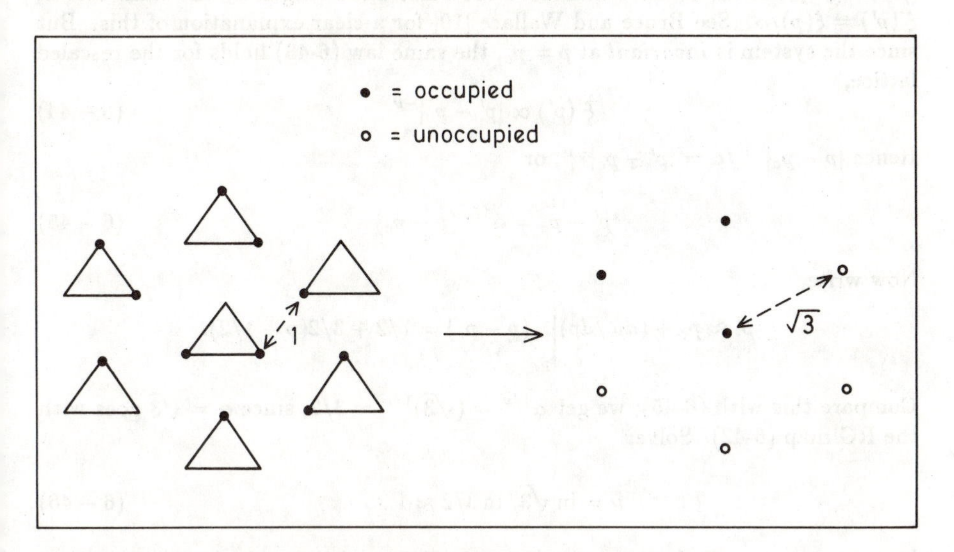

Fig. 6.5 Coarse-grained triangular percolation lattice

The block sites are put at the centers of the original equilateral triangles. Therefore the *scale* (def. lattice spacing) has been increased in the ratio $1 : \sqrt{3}$ (see Fig. 6.5).

If p' is the block site occupation probability, we have

$$p' = R_{\sqrt{3}}(p) \equiv p^3 + 3p^2(1-p) \ . \tag{6-42}$$

> **Ex. 6.1** Verify (6-42). This is a simple exercise in using the laws of probability.

The RG map (6-42) has fixed points $p^* = 0, 1$, and $1/2$. The first two are trivial fixed points. We identify p_c with the third, $p_c = 1/2$, for the reasons given just above. The "flow" under iteration of the RG map (6-42) could be investigated as before, but we are more concerned here with calculating a critical exponent.

We assume that the correlation length $\xi(p)$ has behavior

$$\xi(p) \propto |p - p_c|^{-\nu} \tag{6-43}$$

with critical exponent ν near the critical point. The aim is to calculate ν using the RG method. Now we have increased the scale by a factor $\alpha > 1$ by the coarse-graining operation, so have decreased the correlation length by the same factor, $\xi'(p') = \xi(p)/\alpha$. See Bruce and Wallace [10] for a clear explanation of this. But since the system is *invariant* at $p = p_c$, the same law (6-43) holds for the rescaled lattice,

$$\xi'(p') \propto |p' - p_c|^{-\nu} \quad . \tag{6-44}$$

Hence $|p - p_c|^{-\nu}/\alpha = |p' - p_c|^{-\nu}$, or

$$p' - p_c = \alpha^{1/\nu}(p - p_c) \quad . \tag{6-45}$$

Now write

$$p' \approx p_c + (dp'/dp)\Big|_{p_c}(p - p_c) = 1/2 + 3/2(p - 1/2).$$

Compare this with (6-45); we get $\alpha^{1/\nu} = (\sqrt{3})^{1/\nu} = 3/2$, since $\alpha = \sqrt{3}$ goes with the RG map (6-42). Solve:

$$\nu = \ln\sqrt{3}/\ln 3/2 \approx 1.35 \quad . \tag{6-46}$$

More rigorous methods give $\nu = 4/3$, so the RG result is not bad.

7. PARTIAL DIFFERENTIAL EQUATIONS

7.1 General Remarks

Pde dynamics is the most difficult we have yet met in this survey. One must solve differential equations in several independent variables $x \in \Re^n$ and in time t for fields $\Phi(x,t)$ defined at all points of some region U of $\Re^n$ and some range of times. Thus there are an infinite number of degrees of freedom as opposed to the finite-dimensional phase spaces of flows, sets of N ode's. ("Pde" and "ode" for partial and ordinary differential equations, respectively, are abbreviations common in the literature, and we shall use them.) Add to this the fact that there are different species of pde's (elliptic, hyperbolic, etc.), each with its own kind of initial and boundary conditions in order to be well-posed, while there is only one kind of ode. If the computer is turned loose on a badly posed pde, it prints out reams of nonsense.

Different ways of circumventing these difficulties in pde dynamics have been devised, since the brute force method of simply integrating them from initial and boundary conditions as specified in existence theorems is usually impracticable, even for the largest computers. Solving pde's in physics is as much an art as a science; it takes experience and good intuiton to capture the essential features of a rigorous solution while neglecting a vast number of inessential details. We will not discuss all of these ways here but focus on one method only, the Galerkin method, which is particularly well-suited to revealing those features of a pde dynamics in which nonlinear dynamics ("chaos theory") is interested. It in fact converts the pde dynamics into a *finite-dimensional flow*, about which we know quite a lot (Chap. 2).

7.2 Convection Equations

A typical phenomenon of interest is convection in an incompressible, viscous fluid. The basic equations are the Navier-Stokes equations of hydrodynamics and an energy, or heat, equation. In the Boussinesq approximation and in nondimensionalized form the equations read

$$\partial \mathbf{v}/\partial t + \frac{1}{P}(\mathbf{v}\cdot\nabla)\mathbf{v} = -\nabla p + \Delta\mathbf{v} + RT\hat{\mathbf{n}}\,, \qquad (7-1a)$$

$$P\frac{\partial T}{\partial t} + \mathbf{v}\cdot\nabla T = \Delta T + \hat{\mathbf{n}}\cdot\mathbf{v}\,, \qquad (7-1b)$$

$$\nabla\cdot\mathbf{v} = 0\,, \qquad (7-1c)$$

Boundary conditions. For example with solid walls:

$$\mathbf{v} = 0\,, \quad T = 0\,, \quad \text{a condition on } \frac{\partial T}{\partial n} \quad \text{(on the walls).} \qquad (7-1d)$$

Here $\mathbf{v} \equiv$ fluid velocity, $T \equiv$ absolute temperature, $p =$ pressure, $\Delta \equiv \nabla^2$ is the Laplacian. The *stationary* or *quiescent* solution of the original convection equations is simply

$$\mathbf{v}_0 = 0\,, \nabla T_0 = -A\hat{\mathbf{n}}\,, \quad A > 0 \text{ const.}\,, \qquad (7-2)$$

which describes a zero velocity field and a uniform vertical temperature gradient. $\hat{\mathbf{n}}$ is a unit vector in the direction of the uniform gravitational field $\mathbf{g}$. The inhomogeneous boundary conditions of those original equations for $\mathbf{v}_{tot}$ and T_{tot} specify velocity or its derivatives and temperature or its derivatives on the boundaries in physical 3-space $\Re^3$. Having found the quiescent solution (7-2), one writes pde's for the *excesses* $\mathbf{v}$ and T, defined by

$$\mathbf{v}_{tot} \equiv \mathbf{v}_0 + \mathbf{v}\ (= \mathbf{v})\,, \quad T_{tot} \equiv T_0 + T.$$

Since T_0 and $\mathbf{v}_0$ satisfy the same boundary conditions as T_{tot} and $\mathbf{v}_{tot}$, *the excesses satisfy homogeneous b.c.'s* as in (7-1d), very important for the applicability of the Galerkin method. Let "b.c.'s" be short for "boundary conditions" hereafter; it will save a lot of space. *Homogeneous* b.c.'s mean *linear* b.c.'s: if fields ϕ_1 and ϕ_2 satisfy the b.c.'s, then so does any linear combination $a\phi_1 + b\phi_2$. The excesses and the coordinates are then made dimensionless by dividing them by available dimensional constants (beware, this can be done in several ways!) so that the only residue of the dimensional physical constants of the original convection equations left in Eqs. (7-1) are a few dimensionless constants,

$$R \equiv g\beta A L^4/\nu\chi \equiv \text{Rayleigh number}\,, \quad P \equiv \nu/\chi \equiv \text{Prandtl number.} \qquad (7-3)$$

The reader can refer to any number of good books or articles for this Boussinesq reduction and for the definition of the various physical constants occurring. We follow here Gershuni and Zhukovitskii [17], hereafter GZ. We will only name them here: $\beta \equiv$ coefficient of thermal expansion, $\chi \equiv$ thermal diffusivity, $\nu \equiv$ coefficient of kinematic viscosity, $L \equiv$ characteristic length. $\hat{\mathbf{n}}$, A, and g have already been defined.

7.3 The Rayleigh Problem

If we assume 2D motion in a plane horizontal layer with z pointing up: no y-dependence and $v_y = 0$, Eqs. (7-1) simplify. For a *stream function* $\psi(x, z)$ exists:

$$v_z = \partial\psi/\partial x \equiv \psi_x\ , \quad v_x \equiv -\partial\psi/\partial z \equiv -\psi_z\ , \tag{7-4}$$

which satisfies $\nabla \cdot \mathbf{v} = 0$ identically. (From now on we often use a subscript x or z to mean $\partial/\partial x$ or $\partial/\partial z$.) Now if we take the zx curl of (7-1a), the ∇p term is eliminated. Note that

$$\partial v_x/\partial z - \partial v_z/\partial x = -\psi_{zz} - \psi_{xx} \equiv -\Delta_2\psi\ ,$$

where Δ_2 is the 2D Laplacian. The nonlinear terms in (7-1a) and (7-1b) become Jacobian determinants, defined for functions f and g of x and z as

$$\frac{\partial(f,g)}{\partial(z,x)} \equiv \frac{\partial f}{\partial z}\frac{\partial g}{\partial x} - \frac{\partial f}{\partial x}\frac{\partial g}{\partial z}\ .$$

We then get, where $\dot{f} \equiv \partial f/\partial t$,

$$\Delta_2\dot{\psi} = \Delta_2^2\psi + RT_x + \frac{1}{P}\frac{\partial(\psi, \Delta_2\psi)}{\partial(z,x)}\ , \tag{7-5a}$$

$$P\dot{T} = \Delta_2 T + \psi_x + \frac{\partial(\psi, T)}{\partial(z,x)}\ , \tag{7-5b}$$

$$\text{b.c.'s} : \psi = \psi_{zz} = T = 0 \text{ at } z = 0, 1. \quad (\text{all } x, \text{all } t). \tag{7-5c}$$

Here the homogeneous b.c.'s (7-5c) came from the *Rayleigh problem* b.c.'s: the upper and lower plane boundaries of the layer are *free surfaces* (no tangential fluid stresses), that is,

$$v_z = 0\ , \quad \partial v_x/\partial z = \partial v_y/\partial z = 0$$

on those surfaces. Also L was chosen to be the height of the layer, so that dimensionless $z = 0, 1$ defines those surfaces.

7.4 The Galerkin Method

Equations (7-5) are close to a standard form for nonlinear pde's, which we take as

$$\dot{\Phi} = A\Phi + F(\Phi)\ , \quad \mathcal{L}(\Phi) = 0. \tag{7-6}$$

Here $\Phi(x, t)$, $x \in \Re^n$, is the field or set of p fields in question, thus $\Phi \in \Re^p$. A is a linear operator and F, a nonlinear operator on Φ usually involving partial derivatives with respect to x. $\mathcal{L}(\Phi) = 0$ is the set of homogeneous boundary conditions and possibly linear pde's not involving $\partial/\partial t$, like (7-1c) for example. We shall call $\mathcal{L}(\Phi) = 0$ "the b.c.'s" for short. Now an important concept:

Def. Let $\mathcal{H}$ be the linear (or vector) space of all p-component spatial functions $\phi(x) \in \Re^p$, $x \in \Re^n$, which satisfy $\mathcal{L}(\phi) = 0$. $\quad (7-7)$

(Remember how general a concept "vector space" is: any set of objects closed under taking linear combinations!) Any such $\phi(x)$ is called a *Galerkin vector*. For example, in the Rayleigh system (7-5) any Galerkin vector has the form

$$\phi(x,z) = \begin{pmatrix} \psi(x,z) \\ T(x,z) \end{pmatrix} \in \Re^2 \ ; \ \psi \text{ and } T \text{ satisfy } (7-5c). \qquad (7-8)$$

We assume that $\mathcal{H}$ is *invariant* under A and F, $A : \mathcal{H} \to \mathcal{H}$, $F : \mathcal{H} \to \mathcal{H}$. The last requirement is highly nontrivial.

But we can immediately write down a *complete set* or *basis* $\{\phi_{nk}\}$ of *Galerkin modes* satisfying (7-8):

$$\psi_{nk} \equiv \sin n\pi z \ \sin kx \ , \quad T_{nk} \equiv \sin n\pi z \ \cos kx \ ,$$

$$n = 1, 2, 3, \cdots \quad , \quad 0 \leq k < \infty \qquad (7-9)$$

The basis $\{\phi_{nk}\}$ *spans* the infinite-dimensional vector space $\mathcal{H}$. (One advantage of the Rayleigh b.c.'s (7-5c) is that the Galerkin modes (7-9) take such a simple form. For one free surface and one solid wall or for two solid walls, or for more generally shaped boundaries and any b.c.'s, one cannot in general express the Galerkin basis in terms of finite expressions in elementary functions.)

For the general system (7-6) let $\{\phi_\alpha(x) \mid \text{all } \alpha\}$ be a Galerkin basis for some infinite index set of labels α. Then any solution of (7-6) has the form

$$\Phi(x,t) = \sum_\alpha c_\alpha(t)\phi_\alpha(x) \qquad (7-10)$$

for some functions $c_\alpha(t)$ of time only. Here and below $\sum_\alpha$ may have to be interpreted as an integral $\int d\alpha$ if the labels α are continuous. $\mathcal{H}$ invariant under F, in other words, products of the ϕ_α expressible as linear combinations of all the ϕ_α, was necessary for the truth of (7-10). Write this formally as

$$F\Big(\sum_\alpha c_\alpha\phi_\alpha\Big) = \sum_\alpha f_\alpha(c)\phi_\alpha \quad \text{for any } c \equiv \{c_\alpha | \text{ all } \alpha\}. \qquad (7-11)$$

The Galerkin modes form an *algebra* in this sense. For the linear operator A we write

$$A\sum_\beta c_\beta\phi_\beta = \sum_{\alpha,\beta}(a_{\alpha\beta}c_\beta)\phi_\alpha \ , \qquad (7-12)$$

which introduces the *matrix elements* $a_{\alpha\beta}$ of A. Then when we put (7-10), (7-11), and (7-12) into (7-6) and equate coefficients of the linearly independent ϕ_α to zero, we get the equations of the infinite-dimensional nonlinear *flow*

$$\dot{c}_\alpha = \sum_\beta a_{\alpha\beta} c_\beta + f_\alpha(c) , \quad \text{all } \alpha , \qquad (7-13)$$

that is, an infinite set of coupled nonlinear first order ode's in time for the $c_\alpha(t)$! Eq. (7-13) is exact, and equivalent to the pde system (7-6) in principle.

The idea of the Galerkin method is to approximate the solution (7-10) by using a finite set of the most important modes, say $\phi_1, \phi_2, \cdots \phi_M$ after renumbering the labels α. One writes $\Phi(x,t) \approx \sum_{i=1}^M c_i(t)\phi_i(t)$ and gets an approximate closed set of M ode's by keeping only M of the Eqs. (7-13) and dropping all the c_i, $i > M$, in those equations:

$$\dot{c}_i = \sum_{j=1}^{M} a_{ij} c_j + f_i(c') ,$$

$$c' \equiv \{c_1, c_2, \cdots c_M\} , \quad i = 1,2,3,\cdots M. \qquad (7-14)$$

Choosing the "most important modes" is of course an art. See further remarks below. Technically, one *truncates* the infinite set (7-13) by projecting it onto the finite-dimensional subspace $\mathcal{H}_M \subset \mathcal{H}$ spanned by $\phi_1, \phi_2, \cdots \phi_M$. In particular, $f_i(c')$ comes from $f_i(c)$ by setting to zero all c_i, $i > M$. This prescription is all right if the $f_\alpha(c)$ admit power series expansions in the c_β.

One method of choosing a Galerkin basis of $\mathcal{H}$ has some advantages. We use the *normal modes* $g_i(x)$ of the linearized problem $\dot{\Phi} = A\Phi$, with the same b.c.'s $\mathcal{L}(\Phi) = 0$. Look for solutions $\Phi(x,t) = e^{\lambda t} g(x)$. Then, cancelling $e^{\lambda t}$, we get

$$\lambda g = A g , \; \mathcal{L}(g) = 0 . \qquad (7-15)$$

Let $g_i(x)$, $\lambda_i(x)$, $i = 1,2,3,\cdots$ be the set of eigenfunctions and eigenvalues of A. Usually the set $\{g_i\}$ spans $\mathcal{H}$; if generalized eigenfunctions of A exist (see appendix A), they must be adjoined to $\{g_i\}$ in order to span $\mathcal{H}$. Assuming that the proper eigenfunctions do span $\mathcal{H}$, we can use them instead of the general Galerkin basis used above. The flows (7-13) and (7-14) simplify because the matrix (a_{ij}) is diagonal. The approximate flow then reads

$$\dot{c}_i = \lambda_i c_i + f_i(c') , \; i = 1,2,\cdots M. \qquad (7-16)$$

7.5 Rayleigh Problem by the Galerkin Method

Let us come back to the specific system (7-5). We can put it into standard form (7-6) by dividing (7-5b) by the number P and "dividing" (7-5a) by the operator

Δ_2. The latter is all right if Δ_2 has an inverse Δ_2^{-1} on the space spanned by (7-9); and this is in fact so, since

$$\Delta_2\phi_{nk} = -(n^2\pi^2 + k^2)\phi_{nk} \equiv -\gamma_{nk}\phi_{nk} \,, \qquad (7-17)$$

and γ_{nk} is always positive. So we rewrite

$$\dot{\psi} = \Delta_2\psi + R\Delta_2^{-1}T_x + \frac{1}{P}\Delta_2^{-1}\frac{\partial(\psi,\Delta_2\psi)}{\partial(z,x)} \,, \qquad (7-18a)$$

$$\dot{T} = \frac{1}{P}\Delta_2 T + \frac{1}{P}\psi_x + \frac{1}{P}\frac{\partial(\psi,T)}{\partial(z,x)} \,, \qquad (7-18b)$$

$$\psi = \psi_{zz} = T = 0 \text{ at } z = 0,1 \text{ (all } x \text{ and all } t). \qquad (7-18c)$$

The normal modes are (GZ [17])

$$g_{nk}^{\pm} \equiv \begin{pmatrix}\psi_{nk}^{\pm}\\ T_{nk}^{\pm}\end{pmatrix} = \begin{bmatrix} k^{-1}a_{nk}^{\pm}\sin n\pi z\ \sin kx \\ b_{nk}^{\pm}\sin n\pi z\ \cos kx\end{bmatrix}, \qquad (7-19a)$$

$$\lambda_{nk}^{\pm} = -\frac{P+1}{2P}\gamma_{nk} \pm \left[\left(\frac{P-1}{2P}\right)^2\gamma_{nk}^2 + \frac{Rk^2}{P\gamma_{nk}}\right]^{\frac{1}{2}}, \qquad (7-19b)$$

$$a_{nk}^{\pm} = [\lambda_{nk}^{\pm}P + \gamma_{nk}]\, b_{nk}^{\pm} \,, \ \gamma_{nk} \equiv n^2\pi^2 + k^2. \qquad (7-19c)$$

(Please note: we write time-dependence $e^{\lambda t}$ of normal modes, GZ write $e^{-\lambda t}$. So our eigenvalues (7-19b) are the negatives of GZ's, and we have changed some other notation accordingly.) The complete label i is $(\pm, n, k)$ with $n = 1, 2, 3, \cdots$ and $0 \le k < \infty$. The parameter μ is the pair R, P.

Local bifurcation values are determined by $\lambda_{nk}^{+} = 0$ for each n and k. All $\lambda_{nk}^{-} < 0$. Then the mode $(+, n, k)$ changes from decaying ($\lambda < 0$) to growing ($\lambda > 0$) as λ increases through zero. (Note all λ's are real.) See Fig. 7.1.

These *critical Rayleigh numbers* come out to be

$$R_{nk} = k^{-2}\gamma_{nk}^3 \qquad (7-20)$$

from (7-19b), independent of P. So R_{nk} increases monotonically in n for any k. The smallest critical Rayleigh number $\equiv R_1 = (27/4)\pi^4$ (at $n = 1$ and $k = \pi/\sqrt{2}$). R_1 is thus the threshold for convection; for $R < R_1 \approx 657.5$ there is only the quiescent solution (7-2).

The wave number k in the x-direction can be fixed by various experimental techniques, for example, by the aspect ratio of the cell approximating the infinite plane horizontal layer of the theory (see Chap. 8). We consider it fixed hereafter. Then if $R_{mk} < R < R_{m+1\ k}$, there will be a complicated convection with m

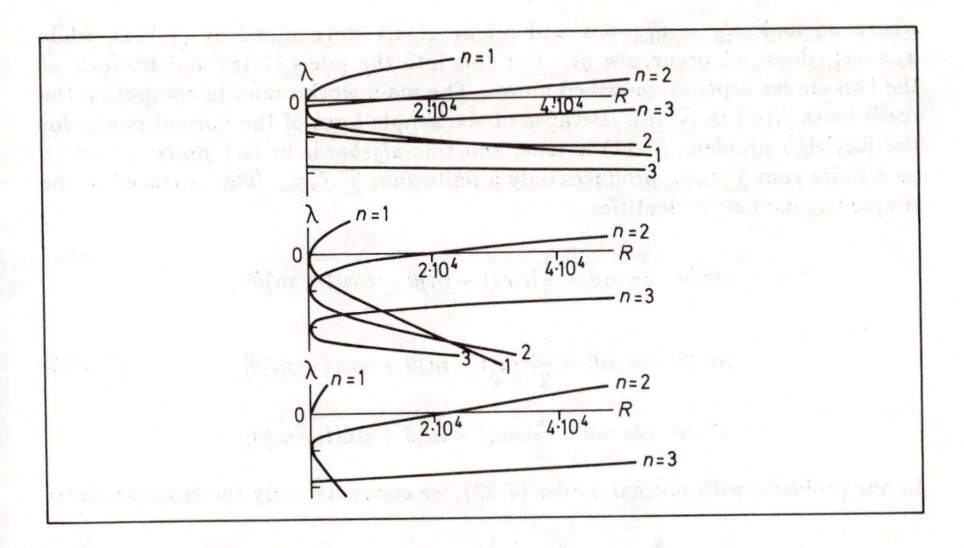

Fig. 7.1 Showing the eigenvalues $\lambda^{\pm}_{nk}$ for $k = 2$ and $n = 1, 2, 3$ as functions of R. Top: $P = 3$, middle: $P = 1$, bottom: $P = \frac{1}{3}$.

growing modes and the rest, decaying modes. Thus it would seem reasonable to try to approximate the nonlinear motion in this regime by a Galerkin approximation including these m modes plus possibly a few others. (The flow is, after all, nonlinear, so we cannot rule out the influence of decaying modes, even as $t \to \infty$.)

Let us fix k and assume $R_{1k} < R < R_{2k}$. Then the unstable subspace E^u will be 1D only, spanned by $g_1 \equiv g^{+}_{1k}$, since $\lambda^{+}_{1k} > 0$. All other modes are decaying, $\in E^s$. If we want a physically good Galerkin approximation with $M > 1$ modes, how do we choose the further modes? There is currently no clear, agreed-upon algorithm for this choice. The pioneering work of Lorenz [28] and succeeding work have provided some hints. For example, $g_2 \equiv g^{+}_{20}$ should certainly be included. We can understand this by noting that the nonlinear terms in (7-18) *generate* this mode from g^{+}_{1k} in the sense of (7-11). Namely, if F acts on the single mode g^{+}_{1k}, it produces the mode g^{+}_{20}. This idea can be made into a precise algorithm for generating a sequence of improving Galerkin approximations (see Ingraham [25]).

So now, as a simple model Galerkin approximation, take

$$\Phi(x,t) = c_1(t)g_1(x) + c_2(t)g_2(x) \,, \tag{7-21}$$

where $x \in \Re^2$ stands for (x, z). From (7-19) the normal modes are

$$g_1 = \begin{pmatrix} k^{-1}a_1 \sin\pi z \, \sin kx \\ \sin\pi z \, \cos kx \end{pmatrix}, \quad g_2 = \begin{pmatrix} 0 \\ \sin 2\pi z \end{pmatrix}, \tag{7-22}$$

where we took $b^+_{1k} = b^+_{20} = 1$ and set $a_1 \equiv a^+_{1k}$ determined by (7-19c), while $a_2 \equiv a^+_{20}$ does not occur, see g_2. Put this into the pde's (7-18) and *truncate* at the two modes kept, as described above. The main work comes in computing the coefficients $f_i(c')$ in (7-16). Because of the simple form of the normal modes for the Rayleigh problem, (7-11) is true, and this *algebra* is in fact *finite*: F acting on a finite sum $\sum c_\alpha \phi_\alpha$ produces only a finite sum $\sum f_\alpha \phi_\alpha$. This is traced to the simple trigonometric identities

$$\sin l\theta \quad \sin m\theta = \frac{1}{2}[\cos(l-m)\theta - \cos(l+m)\theta] \,,$$

$$\cos l\theta \quad \cos m\theta = \frac{1}{2}[\cos(l-m)\theta + \cos(l+m)\theta] \,, \qquad (7-23)$$

$$\sin l\theta \quad \cos m\theta = \frac{1}{2}[\sin(l-m)\theta + \sin(l+m)\theta].$$

In our problem, with normal modes (7-22), we encounter only the cross products

$$\sin \pi z \ \cos 2\pi z = \frac{1}{2}(\sin 3\pi z - \sin \pi z) \,,$$

$$\sin \pi z \ \cos \pi z = \frac{1}{2} \sin 2\pi z \,, \qquad (7-24)$$

as can be verified. Since we keep only the two modes g_1 and g_2, the $\sin 3\pi z$ will not contribute at all. Eventually we arrive at the flow (7-16) for $M = 2$. By solving it numerically we can get some idea of the actual flow near R_{1k}.

Prob. 7.1 Carry through the Galerkin approximation (7-21). For the flow (7-16) you should find

$$\dot{c}_1 = \lambda_1 c_1 + \frac{\pi a_1 a_3}{P(a_3 - a_1)} c_1 c_2 \,, \ \dot{c}_2 = \lambda_2 c_2 - \frac{\pi a_1}{2P} c_1^2 \,, \qquad (7-25)$$

where $a_1 \equiv a^+_{1k}$, $a_3 \equiv a^-_{1k}$ determined by (7-19c) with $b^+_{1k} = b^-_{1k} = 1$ and $\lambda_1 \equiv \lambda^+_{1k} > 0$, $\lambda^+_{20} < 0$.

Prob. 7.2 Carry through the three-normal mode approximation

$$\Phi(x,t) = c_1(t) g_1(x) + c_2(t) g_2(x) + c_3(t) g_3(x)$$

with $g_3 \equiv g^-_{1k}$ and g_1, g_2 as before.

This latter flow in $\Re^3$ is equivalent to the 3-Galerkin mode approximation of Lorenz [28] for $k = \pi/\sqrt{2}$, as can be verified.

7.6 Cautionary Remarks

The short survey of the general Galerkin method in section 7.4 is deceptively simple. The real work is concealed in the Eqs. (7-6) and (7-12). The reduction to normal form (7-6) depends on which fields one wants to isolate; the others are eliminated at the expense of raising the differential order. For example, (7-5a) is fourth order, while the original equations are only second order. When 3D motion is considered, the representation

$$\mathbf{v} = \mathbf{v}_{pol} + \mathbf{v}_{tor} \equiv \nabla \times (\nabla \times \mathbf{r}\psi) + \nabla \times \mathbf{r}\omega \;, \qquad (7-26)$$

which satisfies the incompressible constraint $\nabla \cdot \mathbf{v} = 0$ identically, is often used. This already increases the differential order by two, and further manipulation to separate the equations for the *poloidal* and *toroidal stream functions* ψ and ω as much as possible increases it still further.

As to (7-12), it takes considerable stamina to do the algebra which results in the explicit coefficients $f_\alpha(c)$ when M is not small. Further, this algebra need not be *finite*: finite number of terms on the right of (7-11) when a finite linear combination of modes is the input on the left. This infinite sum may also involve integrals. This, unfortunately, is the case usually encountered – the Rayleigh problem has an unusual simplicity because of the simple, finite closure properties of trigonometric functions under multiplication. In fact, the Galerkin method with m growing modes is only practicable when the nonlinear operator F generates a few further modes with nonnegligible coefficients $f_\alpha(c)$. It is hard to say *a priori* which pde systems and b.c.'s guarantee this criterion. But the reader should not forget that we are confronting real nonlinear motion laws head on by the Galerkin method. Neat, simple solutions cannot be expected.

Working out probs. 7.1 and 7.2 will give the serious reader an understanding of the general Galerkin method. There is no substitute for the actual labor of applying the nonlinear operator to a sum of normal modes, producing a general Galerkin vector, then reëxpressing this as a sum of normal modes.

7.7 Patterns of Convection

We look at the pattern of convection for our simple Rayleigh problem. By (7-4)

$$v_z = \psi_x \;, \; v_x = -\psi_z \;. \qquad (7-27)$$

Since we have 2D motion with no y-dependence, the convection must be in the form of "y-rolls," with axes parallel to the y-axis. For a single mode we get

$$v_z \propto \sin n\pi z \cos kx \;, v_x \propto -k^{-1} n\pi \cos n\pi z \;\sin kx \qquad (7-28)$$

from (7-19a). The velocity *nodes* ($\mathbf{v} = 0$) are at the lattice points.

$$(kx, n\pi z) = (p\,\pi/2,\; q\,\pi/2)\;;\; p, q \text{ integers, both even or both odd.} \qquad (7-29)$$

Further, the stability of these velocity fixed points is

$$p, q \text{ both odd} \Rightarrow \text{center}\ ; \quad p, q \text{ both even} \Rightarrow \text{saddle point.} \qquad (7-30)$$

This is most easily seen by drawing velocity vectors along the points of the horizontal and vertical lines through lattice points (7-29), using (7-28). The streamlines of these rolls fill out the squares, they are not circles around the centers. The boundaries of the squares are streamlines, the separatrices of the saddle points. See Fig. 7.2.

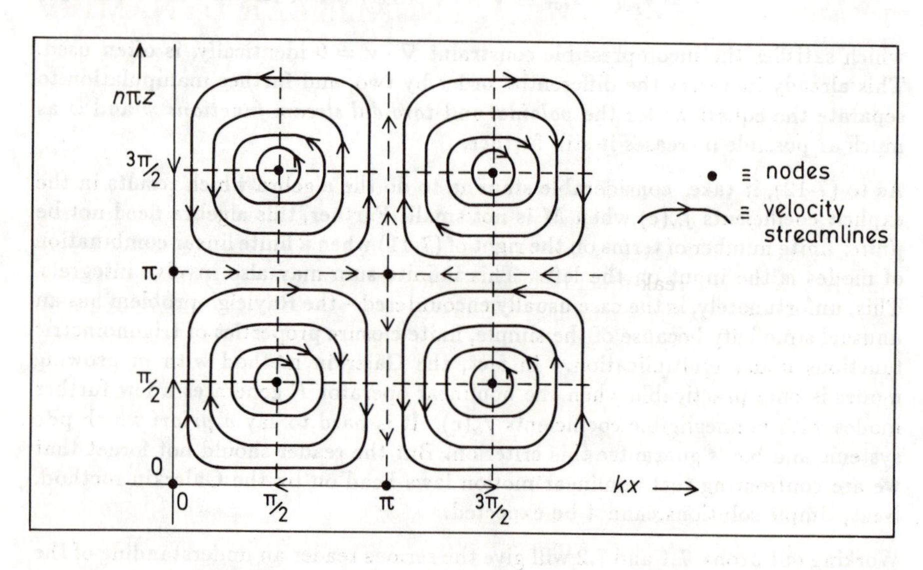

Fig. 7.2 Convection roll fluid flow patterns

8. EXPERIMENTAL REALIZATIONS OF NONLINEAR DYNAMICS

8.1 Introduction

So far we have seen several theoretical "paths to chaos." There is the period-doubling route 2^n-cycles $\to$ "chaos" as $n \to \infty$ in 1D maps, caused by increasing the parameter. Then there is the route n-tori $\to$ KAM tori plus stochastic layers ("chaos") from the break-up of resonance zones in Hamiltonian dynamics. This is caused by increasing the strength of a nonintegrable perturbation to an integrable Hamiltonian. Then there is the intermittency route: periodic motion interrupted by bursts of "chaotic" behavior, which approaches a fully "chaotic" regime as some parameter is varied. All of these routes have been seen experimentally. We describe a few of these experiments briefly in this chapter.

We shall discuss pde systems in the rest of this section, focusing on incompressible viscous fluid flow. To see the various stable regimes, the bifurcations connecting them, the associated universal numbers, etc., it is usually sufficient to measure any of the physical fields, say v_x, v_y, v_z, or T, at a fixed point in space as a function of time.

This brings up the question: what is the connection of the actual fluid velocity and temperature distributions in 3D configuration space and the mathematical image of it, the M functions of time $c_1(t)$, $c_2(t)$, $\cdots c_M(t)$ which we study in a Galerkin approximation? This is determined of course by the expansion

$$\Phi(x,t) = \sum_i c_i(t) g_i(t) \tag{8-1}$$

of the fields in normal modes g_i. From that we can understand the following correspondences.

a. A stable fixed point c_i^* of the *flow*

$$\dot{c}_i = \lambda_i c_i + f_i(c)\ , \tag{8-2}$$

$\dot{c}_i^* = 0$, all i, corresponds to a stable stationary fluid motion (*and* temperature distribution understood hereafter), typically convection rolls.

b. If the flow (8-2) shows a stable periodic orbit (limit cycle) $c_i(t+T) = c_i(t)$, all t, then the fluid motion becomes periodic with period T, a regime called *oscillatory*. (A critical *Reynolds number* $R_e \equiv vL/\nu$ governs the onset of oscillatory motion. Here $v \equiv$ characteristic velocity, and L and ν were defined in chap. 7.) Period-doubling in the flow at a bifurcation value of the Rayleigh number leads to a bifurcation in the fluid motion to an oscillatory motion of period 2T. If, finally, the flow enters a "chaotic" regime, say it is *chaotic* in the precise sense of definition (5-5), the fluid motion is correspondingly chaotic. One could in principle measure the sensitive dependence of the fluid quantities, say $T(\mathbf{r}, t)$, by varying the pde initial conditions slightly.

c. On fixing the wave number k by experimental techniques, the following is presumably the basic idea. Experiments use small rectangular parallelopipedal cells to approximate the infinite horizontal layer of theory (section 7.3). See Fig. 8.1.

First, $L_x >> L_y$ is desirable. Second, if an integral number l of identical convection rolls form, one has $kL_x = l\pi$ (see Fig. 7.2), which fixes k.

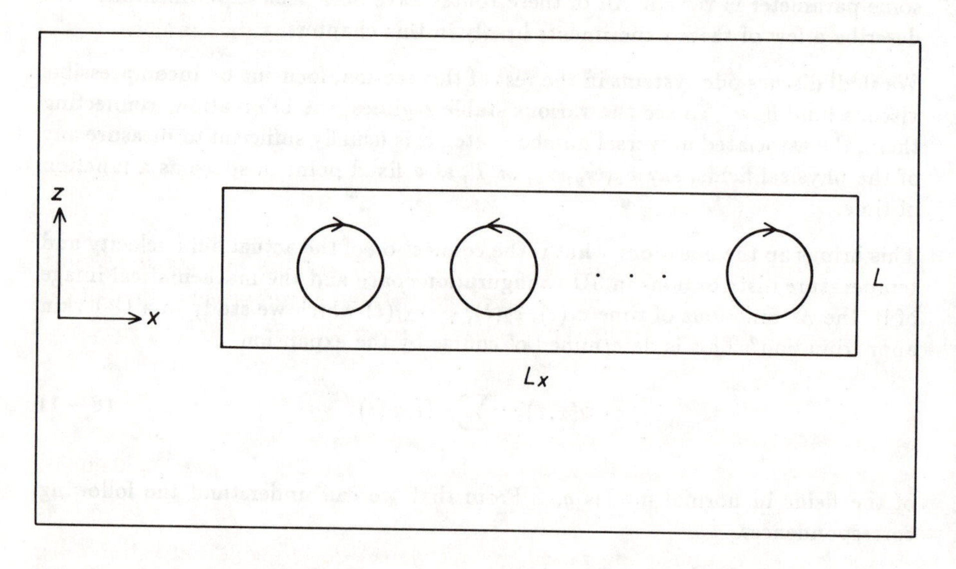

Fig. 8.1 Experimental "rectangular" cell of height L and x-width L_x

8.2 Some Specific Experiments

8.2.1 PDE Systems

a. Libchaber and Maurer [27]. Liquid ^{4}He in a small cell, whose top and bottom were highly conducting plates. Two bolometers imbedded in the upper plate give local temperature readings. The cell dimension "favored" two or three convection rolls. The Prandtl number was small, $0 < P < 1$. Temperatures were in the range 2.5–4.5K. The temperature difference $\Delta T_0 = AL$ (cf. (7-2)) was only the order of mK for the onset of convection.

As the Rayleigh number was increased above the regime of stable stationary convection, they saw: first, oscillatory motion of a frequency ν_1 (which increased linearly with R), then the appearance of a second, incommensurable frequency ν_2, then frequency locking $\nu_2 \to \nu_1/2$, then a cascade of subharmonics (period-doubling) $\nu_1/2, \nu_1/4, \nu_1/8, \nu_1/16, \cdots$ resolved up to $\nu_1/16$. These are all obtained from the signal $T_1(t)$ of one of the bolometers. Other regimes were also seen, which we skip here.

Frequency locking is a phenomenon in nonlinear oscillators whereby for two close enough frequencies ν_1, ν_2 of the motion ν_2 suddenly jumps to a subharmonic of ν_1. The reader is referred to Bak [6].

Let R_n be the period-doubling bifurcation values of the Rayleigh number. It was found that

$$\tilde{\delta} \equiv \lim_{n\to\infty} \frac{R_n - R_{n-1}}{R_{n+1} - R_n} \approx 3.5 \qquad (8-3)$$

to the observed resolution. Cf. $\delta = 4.6692\cdots$ for the corresponding logistic map ratio.

b. Libchaber, Laroche, and Fauve [26]. Liquid mercury in a small cell with top and bottom plates of copper. A single bolometer was imbedded in the bottom plate. The aspect ratio of the cell "favored" four convection rolls. The rolls were stabilized by a uniform magnetic field parallel to their axes. (We remark that this goes beyond the system (7-1) into MHD, or magnetohydrodynamics. MHD is the proper dynamics for some extremely important and interesting areas, for example, plasma physics, stellar interiors, the geodynamo (source of the Earth's magnetic field).)

As the Rayleigh number was raised beyond the regime of stable convection rolls, oscillatory motion set in. Then a subharmonic cascade $\nu_1, \nu_1/2, \nu_1/4, \nu_1/8, \nu_1/16$ was seen in the reading $T(t)$ of the bolometer. See Fig. 8.2 for the time series

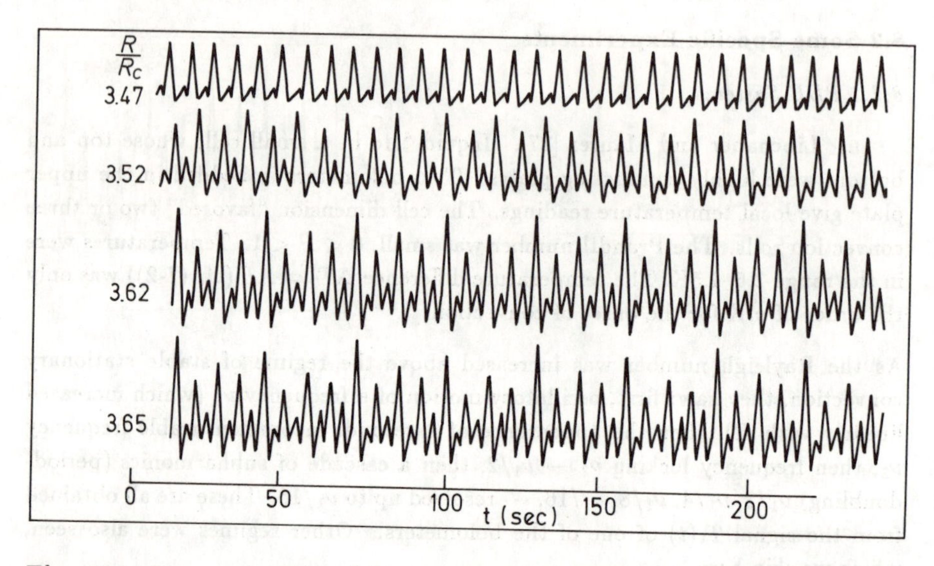

Fig. 8.2a Time series $T(t)$ of the temperature for various Rayleigh numbers R. R_c is our $R_1 \equiv$ absolute minimum of the R_{nk} (see chap. 7) (From Libchaber, Laroche, and Fauve [26], with permission)

$T(t)$ and its power spectrum. The number $\tilde{\delta} \equiv$ (8-3) was observed to be ≈ 4.4, nearer the universal value of the logistic map.

For the theory of the complicated set of stable regimes and their bifurcations, see Busse [11]. For the experiments summarized in this chapter, see Cvitanović [14] (part 2). From now on our descriptions will be briefer.

c. Gollub and Swinney [19]. They measured a fluid velocity component at a fixed point in a *rotating* fluid. The system is an incompressible viscous fluid, but is not exactly that of the convective system (7-1).

d. Bergé *et al.* [8]. They measured v_z locally and an averaged $\nabla_z T$ in silicone oil, where the Prandtl number P is large, $P \approx 130$. They found evidence of intermittency (section 3.10.3).

8.2.2 Other Systems

e. Hudson and Mankin [23]. A chemical system, the Belousov-Zhabotinskii reaction. They measured the concentration of one chemical species by an indirect method. The system here is an actual finite-dimensional flow, cf. the simple Brusselator model of section 2.7.1.

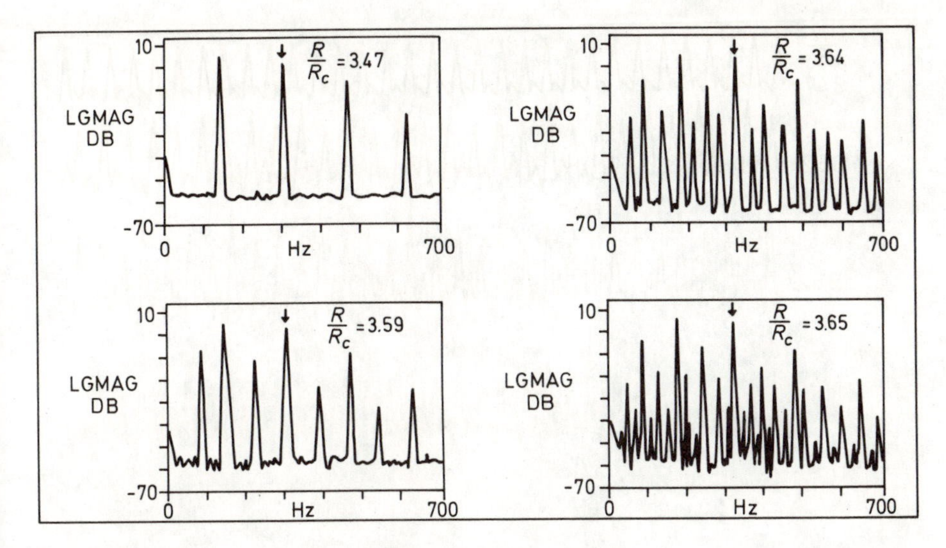

Fig. 8.2b Power spectrum of $T(t)$. Arrows indicate the peak at ν_1. (From Libchaber, Laroche, and Fauve [26], with permission)

f. Arecchi *et al.* [2]. A quantum optical system, a periodically perturbed CO_2 laser. Mathematically, the system is a nonautonomous flow in $\Re^2$ (thus an autonomous system in one higher dimension).

g. Testa *et al.* [34]. An electronic system, a driven nonlinear oscillator. This is a nonautonomous flow in $\Re^2$. Electronic systems with their precise controls and measurements, allow the cleanest tests of the theory.

h. Guevara, Glass, and Shrier [21]. Living matter system (!), cardiac cells perturbed by electrical current pulses.

REFERENCES

1. V. M. Alekseev and M. V. Yacobson, *Phys. Repts.* **75** (1981) 291. See especially pp. 291-296.

2. F. T. Arecchi, R. Meucci, G. Puccioni, and J. Tredicce, *Phys. Rev. Lett.* **49** (1982) 1217.

3. V. I. Arnold and A. Avez, *Ergodic Problems of Classical Mechanics* (Benjamin, New York, 1968).

4. V. I. Arnold, *Mathematical Methods of Classical Mechanics* (Springer, New York, 1978).

5. H. Bai-Lin, *Chaos* (World Scientific, Singapore, 1984).

6. P. Bak, "The Devil's Staircase," *Phys. Today* **Dec** (1986) 39.

7. G. Benettin, L. Galgani, A. Giorgilli, and J. M. Strelcyn, *Meccanica* **15** (1980) 9.

8. P. Bergé, M. Dubois, P. Manneville, and Y. Pomeau, *J. Phys. Lett.* **41** (1980) L341.

9. P. Bergé, Y. Pomeau, and C. Vidal, *Order Within Chaos* (Wiley, New York, 1984).

10. A. Bruce and D. Wallace, in *The New Physics*, ed. P. Davies (Cambridge University Press, New York, 1989) Chap. 8.

11. F. H. Busse, *Rept. Progr. Phys.* **41** (1978) 1929.

12. D. Chandler, *Introduction to Modern Statistical Mechanics* (Oxford University Press, New York, 1987).

13. B. V. Chirikov, *Phys. Rept.* **52** (1979) 263.

14. P. Cvitanović, *Universality in Chaos* (Adam Hilger, Bristol, 1984).

15. R. L. Devaney, *An Introduction to Chaotic Dynamical Systems* (Benjamin/Cummings, Menlo Park, California, 1986).

16. M. J. Feigenbaum, *J. Stat. Phys.* **19** (1978) 25.

17. G. Gershuni and E. Zhukovitskii, *Convective Stability of Incompressible Fluids* (Keter, Jerusalem, 1976).

18. H. Goldstein, *Classical Mechanics 2nd. ed.* (Addison-Wesley, Reading, Massachusetts, 1980).

19. J. Gollub and H. Swinney, *Phys. Rev. Lett.* **35** (1975) 927.

20. J. Guckenheimer and P. Holmes, *Nonlinear Oscillations, Dynamical Systems, and Bifurcations of Vector Fields* (Springer, New York, 1983).

21. M. Guevara, L. Glass, and A. Shrier, *Science* **214** (1981) 1350.

22. M. Hénon and C. Heiles, *Astron. J.* **69** (1964) 73.

23. J. L. Hudson and J. C. Mankin, *J. Chem. Phys.* **74** (1981) 6171.

24. R. L. Ingraham, Physics Dept. Preprint (New Mexico State University, 1988).

25. R. L. Ingraham, SIAM *J. Appl. Math.* **50** (1990) 956.

26. A. Libchaber, C. Laroche, and S. Fauve, *J. Phys. Lett.* **43** (1982) L211.

27. A. Libchaber and M. Maurer, in *Nonlinear Phenomena at Phase Transitions and Instabilities*, ed. T. Riste (Plenum, New York, 1982) p. 259.

28. E. Lorenz, *J. Atmos. Sci.* **20** (1963) 130.

29. H. Maris and L. Kadanoff, *Am. J. Phys.* **46** (1978) 652.

30. P. W. Milonni, M.- L. Shih, and J. R. Ackerhalt, *Chaos in Laser-Matter Interactions* (World Scientific, Singapore, 1987).

31. M. Plischke and B. Bergersen, *Equilibrium Statistical Mechanics* (Prentice Hall, Englewood Cliffs, New Jersey, 1989).

32. Y. Pomeau, B. Dorizzi, and B. Grammatikos, *Phys. Rev. Lett.* **56** (1986) 681.

33. F. Reif, *Fundamentals of Statistical and Thermal Physics* (McGraw-Hill, New York, 1965).

34. J. Testa, J. Pérez, and C. Jeffries, *Phys. Rev. Lett.* **48** (1982) 714.

35. G. H. Walker and J. Ford, *Phys. Rev.* **188** (1969) 416.

36. S. Wiggins, *Introduction to Applied Nonlinear Dynamical Systems and Chaos* (Springer, New York, 1990).

37. L. S. Young, *IEEE Transactions on Circuits and Systems* **30** (1983) 599.

APPENDIX A

GENERAL LINEAR STABILITY ANALYSIS

We consider the linear flow $\dot{x} = Ax$, $x \in \Re^N$, $A \equiv$ real $N \times N$ matrix. In the general case the N proper eigenvectors e_i of A, see (2-4), are not linearly independent ($\equiv$ do not span $\Re^N$). A simple example in $\Re^2$ is $\binom{1\ 1}{0\ 1}$. The double eigenvalue $\lambda = 1$ has only one linearly independent eigenvector $\binom{1}{0}$. However, in any case we can introduce the generalized eigenvectors, which *do* span $\Re^N$. The vector f is called a *generalized eigenvector* of A to eigenvalue λ if

$$(A - \lambda \mathbf{1})^p f = 0 \text{ for some integer } p,\ 1 \leq p \leq N. \qquad (A-1)$$

For $p = 1$, f reduces to an ordinary, or proper, eigenvector.

Let f_i, $i = 1, 2, \cdots N$, be the linearly independent generalized eigenvectors. For matrix $A = (a_{ij})$, say they have components f_{ij}, that is, f_{ij} is the jth component of vector f_i. Then A can be put into Jordan canonical form by the similarity transformation $T^{-1}AT$, by using the matrix T whose ijth matrix element is $T_{ij} \equiv f_{ij}$. That is, the vector f_i forms the ith column of the matrix T. If the f_i are properly ordered, A takes *Jordan canonical form*, namely A is in *block form*

$$A = \begin{bmatrix} B_1 & & & \bigcirc \\ & B_2 & & \\ & & \ddots & \\ \bigcirc & & & B_q \end{bmatrix} \qquad (A-2)$$

with submatrices B_k along the diagonal, where each B_k has the form

$$B = \begin{bmatrix} \lambda & 1 & & & \bigcirc \\ & \lambda & 1 & & \\ & & \ddots & \ddots & \\ & & & \lambda & 1 \\ \bigcirc & & & & \lambda \end{bmatrix}. \qquad (A-3)$$

That is, a single eigenvalue λ on the main diagonal, 1's along the main superdiagonal, and zeros elsewhere. The reader can easily check that this includes the case of A diagonalizable. Then the B_k are all 1×1, simply the eigenvalues λ_k of A. It is also clear that if all the N eigenvalues λ_i are different, A must be diagonalizable.

To test your understanding of this, try to do the following exercises.

Ex. A.1

a. Given a generalized eigenvector f, let p be the minimal exponent for which (A-1) holds. Prove that the vectors f, $(A - \lambda\mathbf{1})f$, $(A - \lambda\mathbf{1})^2 f$, $(A - \lambda\mathbf{1})^{p-1} f$ are linearly independent.

b. If $p = N$ in part (a), we have N linearly independent vectors $e_i \equiv (A - \lambda\mathbf{1})^i f$, $i = 0, 1, 2, \cdots N - 1$. Show that in this case, if the e_i are taken as basis elements, A has the form (A-4) of a single Jordan block. (Hint: remember the relation

$$Ae_i = \sum_{j=1}^{N} a_{ji} e_j \ , \ i = 1, 2, \cdots N$$

defining the matrix elements of a_{ji} of a linear operator A.)

APPENDIX B

PROOF OF NO SENSITIVE DEPENDENCE IN THE QUANTUM-MECHANICAL SELF-CORRELATION

This map of the state vector is defined

$$C(t,\tau) \equiv \langle\psi(t)|\psi(t+\tau)\rangle \ , \ t \equiv \text{ time} \ , \ \tau \text{ a const.} \qquad (B-1)$$

This is a complex number; the corresponding metric $\tilde{d}$ is the absolute value of the difference. Thus, given $\|\psi' - \psi\| < \delta$ small, we want to prove

$$\Delta C \equiv \Big|\langle\psi'(t)|\psi'(t+\tau)\rangle - \langle\psi(t)|\psi(t+\tau)\rangle\Big| \qquad (B-2)$$

always small, where $\psi(0) \equiv \psi$, $\psi'(0) \equiv \psi'$, and both are normalized.

Now $\psi(t) = \mathsf{U}(t,0)\ \psi(0)$, $\psi(t+\tau) = \mathsf{U}(t+\tau,0)\ \psi(0)$, and similarly for $\psi'(t)$ and $\psi'(t+\tau)$, where U is the unitary time-evolution operator. Here we place no restriction on the QM system : $\mathcal{H}$ has any dimension, including ∞; the energy operator is any whatsoever, and may depend explicitly on t. We can thus rewrite

$$\langle\psi'(t)|\psi'(t+\tau)\rangle = \langle\psi'|\widetilde{\mathsf{U}}(t,\tau)|\psi'\rangle \ , \qquad (B-3)$$

where $\widetilde{\mathsf{U}}(t,\tau) \equiv \mathsf{U}^{-1}(t)\ \mathsf{U}\,(t+\tau)$, and the same for ψ.

Now we need a convenient representation for *any* ψ' close to ψ, such that $\|\psi' - \psi\| = \delta_1$ say, $0 < \delta_1 < \delta$. This is

$$\psi' = N^{-1}(\psi + \delta_1\chi), \text{ where } \|\chi\| = 1, \ \langle\chi|\psi\rangle = 0 \ ,$$

$$N \equiv (1+\delta_1^2)^{\frac{1}{2}} \ , \qquad (B-4)$$

as the reader can verify. We repeat: *any* state ψ' in the δ-neighborhood $\|\psi' - \psi\| < \delta$ can be represented as in (B-4) for some unit state χ normal to ψ and some $\delta_1 < \delta$.

Put (B-4) into (B-3). We get

$$\langle\psi'|\tilde{U}|\psi'\rangle = N^{-2}\left[\langle\psi|\tilde{U}|\psi\rangle + \delta_1^2\langle\chi|\tilde{U}|\chi\rangle + 2\delta_1\ Re\langle\chi|\tilde{U}|\psi\rangle\right]. \qquad (B-5)$$

This is then substituted into ΔC, Eq. (B-2). Now we use some elementary inequalities on complex numbers and the Schwarz inequality in QM:

$$\left|\langle a|b\rangle\right| \leq \|a\| \cdot \|b\|\,, \qquad (B-6)$$

where $\langle a|b\rangle$ is the scalar product of two vectors in $\mathcal{H}$ and $\|a\|^2 \equiv \langle a|a\rangle$, etc. We get

$$\Delta C = \left|(N^{-2}-1)\langle\psi|\tilde{U}|\psi\rangle + N^{-2}\delta_1^2\langle\chi|\tilde{U}|\chi\rangle + 2N^{-2}\delta_1 Re\langle\chi|\tilde{U}|\psi\rangle\right|$$

$$\leq (1-N^{-2})\left|\langle\psi|\tilde{U}|\psi\rangle\right| + N^{-2}\delta_1^2\left|\langle\chi|\tilde{U}|\chi\rangle\right| + 2N^{-2}\delta_1\left|\langle\chi|\tilde{U}|\psi\rangle\right| \leq 1-N^{-2}$$

$$+N^{-2}\delta_1^2 + 2N^{-2}\delta_1 = \frac{2\delta_1(1+\delta_1)}{1+\delta_1^2} < 2\delta(1+\delta)\,, \text{all } t \text{ (and all } \tau). \qquad (B-7)$$

Here the Schwarz inequality gave $\left|\langle\chi|\tilde{U}|\psi\rangle\right| \leq \|\chi\| \cdot \|\tilde{U}\psi\| = 1\cdot 1 = 1$ since $\|\chi\| = \|\psi\| = 1$ and the unitary $\tilde{U}$ does not change the norm. Similarly for the other terms.

Now the formal proof that $C(t,\tau)$ is Lyapunov stable at any initial state ψ, thus has no SD (5-2), is obvious, but let us carry it through to the end for clarity. Refer to Def. (5-6). Given an $\epsilon > 0$. Then for $\|\psi' - \psi\| < \delta(\epsilon)$, where we take $\delta(\epsilon)$ to satisfy

$$2\delta(1+\delta) < \epsilon\,, \qquad (B-8)$$

we have $\Delta C < 2\delta(1+\delta) < \epsilon$, all t, Q.E.D.

GLOSSARY OF PRINCIPAL TERMS

The terms marked with * are defined for both flows (continuous time systems) and map dynamics (discrete time systems). In this glossary, only the definitions for flows are given. The emendations needed to make the definitions apply to discrete time systems can be found by consulting Chapter 3, especially sections 3.2 – 3.8.

algorithmic complexity (AC) – The algorithmic complexity $K(s_N)$ of the finite word s_N is the length (number of symbols) of the shortest algorithm which will print the word.

analytic function – A real-valued function f of real variables $x \in \Re^N$ is analytic in a neighborhood $V \subset \Re^N$ if $f(x)$ admits a power series expansion in x about every point $x_0 \in V$.

asymptotically stable* – The fixed point x^* is asymptotically stable if it is stable and for every neighborhood $V \subset U$ of x^* a neighborhood $V_1 \subset V$ of x^* exists such that $\phi_t(x_0) \to x^*, t \to +\infty$, for every $x_0 \in V_1$.

attracting set – A closed and invariant set Λ is called an attracting set if $\exists$ a neighborhood V of Λ such that $\phi_t(x) \in V$ for $t \geq 0$ and $\phi_t(x) \to \Lambda$, $t \to +\infty$, for all $x \in V$. This definition can be strengthened by requiring some extra properties such as indecomposability (Λ contains a dense orbit), generalized dimension in some range, "chaotic" flow, etc., and then such sets are called attractors or strange attractors.

attractor – see **attracting set.**

autonomous flow – If the defining equations of a flow $\dot{x} = f(x,t) \equiv f(x)$ do not depend explicitly on t, the flow is said to be autonomous.

bifurcation* – The sudden qualitative change in a flow caused by an arbitrarily small change in its velocity field $f(x)$, that is, such that the flows before and after the bifurcation cannot be continuously deformed into each other. More precisely, the parameter value μ_b of the flow $\dot{x} = f_\mu(x)$ is called a bifurcation value if the flow is *not* structurally stable at $\mu = \mu_b$. This means that there exist arbitrarily small perturbations $\delta_1 f(x)$, $\delta_2 f(x)$ of $f_{\mu_b}(x)$ such that the perturbed

flow $\dot{x} = f_{\mu_b}(x) + \delta_1 f(x)$ **is not topologically equivalent to the perturbed flow** $\dot{x} = f_{\mu_b}(x) + \delta_2 f(x)$.

cascade – See **map dynamics**.

center* – A fixed point x^* is a center if it is stable but not asymptotically stable.

center subspace $(E^c)^*$ – A center subspace is the subspace of the phase space for a linear autonomous flow $\dot{x} = Ax$ (where A is a real $N \times N$ matrix), which is spanned by the subset $\{w_1, w_2, \cdots w_{N_c}\}$of eigenvectors of A such that $Re\lambda_i = 0$.

dissipative system – An autonomous flow $\dot{x} = f(x)$ in $\Re^N$ is called dissipative if $\nabla \cdot f \equiv \sum_{i=1}^N \frac{\partial f_i}{\partial x_i} < 0$ there. Since $f(x)$ is the velocity field of the flow in phase space, this negative divergence implies that comoving volume elements are shrinking in time. Hence a nonzero volume of $\Re^N$ asymptotically shrinks to volume 0 under the flow.

finite word – A finite word s_N is a string of N letters from some alphabet $\mathcal{L}$, for example $\mathcal{L} = \{1, 2, \cdots m\}$. A special case is $\mathcal{L} = \{0, 1\}$, the binary alphabet.

fixed point $(x^*)^*$ – A fixed point x^* of a flow $\dot{x} = f(x)$ is defined by $\dot{x}^* = f(x^*) = 0$.

flow – A system of N first order ordinary differential equations in time t,

$$\dot{x} = f(x, t), \quad \dot{x} \equiv \frac{dx}{dt}, \quad x \in \Re^N,$$

defines a flow.

Hamilton-Jacobi (H-J) equation – The Hamilton-Jacobi equation is the single partial differential equation

$$H(q, \partial W / \partial q) = E = \text{const.},$$

for Hamilton's characteristic function W.

Hamiltonian system – The flow

$$\dot{q} = \partial H / \partial p, \quad \dot{p} = -\partial H / \partial q, \quad q, p \in \Re^n,$$

in $\Re^N$, $N \equiv 2n$, is a Hamiltonian system of n degrees of freedom. $H \equiv H(q, p, t)$ is called the Hamiltonian. If H does not explicitly depend on t, the flow is autonomous, and $H(q(t), p(t))$ is a constant (or integral) of the motion. In other words, H is conserved.

homeomorphism – A homeomorphism $h : U \to V$ maps open set U into open set V such that h is continuous and h^{-1} exists and is continuous.

hyperbolic* – A fixed point x^* is called hyperbolic (sometimes nondegenerate) if $Df(x^*)$ has no eigenvalue λ with $Re\lambda = 0$, where $Df(x^*)$ is the Jacobian matrix evaluated at the fixed point. That is, if $E^c = \mathbf{0}$, the zero vector.

integrable Hamiltonian system – There are three ways to define an integrable Hamiltonian system:

1) A Hamiltonian system is integrable if the solution curves $\phi_t(x_0) \equiv (q(q_0,p_0,t), p(q_0,p_0,t))$ are analytic in $x_0 = (q_0,p_0)$ and in t.

2) A Hamiltonian system is integrable if there exist n independent analytic constants of the motion $F_i(q(t),p(t)) = \text{const.}$, $i = 1,2,\cdots n$.

3) A Hamiltonian system is integrable if its Hamilton-Jacobi equation admits a complete solution.

invariant set – Set S is invariant if $\phi_t(S) \subset S$, $-\infty < t < \infty$.

local bifurcation* – A local bifurcation at the fixed point x^* of the flow $f_\mu(x)$ occurs when an eigenvalue λ of the Jacobian matrix $Df_\mu(x^*)$ crosses the imaginary axis in the complex eigenvalue plane. If $Re\lambda(\mu_b) = 0$ and this zero is isolated, μ_b is called a local bifurcation value.

manifold – A manifold is a "continuous" set like a curve, surface, volume, etc., or technically: locally just like $\Re^M$ for some positive integer M.

map dynamics – Let F be a map from $\Re^N$ into $\Re^N$ which is continuous with continuous first derivatives, written $F : \Re^N \rightarrow \Re^N$, $F \in C^1$. F determines a dynamics in which time is discrete by iteration, namely

$$x_{n+1} = F(x_n) \quad , \quad n = \text{integer} \quad .$$

A map dynamics is also called a cascade.

phase space* – The space of all states of a system. For a flow in $\Re^N$, phase space is $\Re^N$ or that subset $U \subset \Re^N$ in which the flow is confined.

saddle point* – A hyperbolic fixed point x^* is called a saddle point if the stable subspace E^s and the unstable subspace E^u of the phase space are both $\neq \mathbf{0}$, the zero vector.

sensitive dependence (SD) on initial conditions – The dynamics D with orbit function ϕ_t and metric $d(x',x)$ has sensitive dependence on initial conditions at state x if there exists an $\epsilon > 0$ such that, for any neighborhood $\mathcal{N}$ of x there exists an $x' \in \mathcal{N}$ and a $t > 0$ such that $d(\phi_t(x'),\phi_t(x)) > \epsilon$.

sink* – A fixed point x^* is called a sink if it is asymptotically stable.

source* – A hyperbolic fixed point x^* is called a source if the stable subspace E^s of the phase space $= \mathbf{0}$, the zero vector.

stable fixed point* – A fixed point x^* is stable in U if for every neighborhood $V \subset U$ of x^* there is a neighborhood $V_1 \subset V$ of x^* such that every solution $x(t) = \phi_t(x_0)$ with $x_0 \in V_1$ is defined and $\in V$ for all $t > 0$.

stable manifold* (W^s) – The stable manifold W^s of a hyperbolic fixed point x^* is the set of all those points which converge to x^* under the flow. That is, all $x \in$ phase space such that $\phi_t(x) \to x^*$, $t \to +\infty$.

stable subspace* E^s – A stable subspace E^s is the subspace of the phase space for a linear autonomous flow $\dot{x} = Ax$ (where A is a real $N \times N$ matrix), which is spanned by the subset $\{v_1, v_2, \cdots v_{N_s}\}$ of eigenvectors of A such that $Re\lambda_i < 0$.

strange attractor – see **attracting set.**

structurally stable – see **bifurcation value.**

topologically equivalent* – The f-flow and the g-flow in sets X and Y respectively are topologically equivalent if and only if there exists a homeomorphism $h : X \to Y$ such that for every t_1, $h \circ \phi^f_{t_1} = \phi^g_{t_2} \circ h$ for some t_2.

topologically transitive – For a dynamics D with orbit function ϕ_t, V is topologically transitive if, for any pair of open sets $U_1, U_2 \subset V$, the image of U_1 under the dynamics intersects U_2 at some time: $\phi_t(U_1) \cap U_2 \neq \emptyset$for some time $t > 0$.

unstable fixed point* – The fixed point x^* is unstable in U if it is not stable, that is, if $\exists$ a neighborhood $V \subset U$ of x^* such that for all neighborhoods $V_1 \subset V$ of x^* $\exists$ an orbit based at $x_0 \in V_1$ which $\notin V$ for some $t > 0$.

unstable manifold* (W^u) – The unstable manifold W^u of x^* is the set of all those points which "diverge from x^* under the flow" or, more precisely, all those points which converge to x^* as time runs backwards: $\phi_t(x) \to x^*$, $t \to -\infty$.

unstable subspace* E^u – An unstable subspace E^u is the subspace of the phase space for a linear autonomous flow $\dot{x} = Ax$ (where A is a real $N \times N$ matrix), which is spanned by the subset $\{u_1, u_2, \cdots u_{N_u}\}$ of eigenvectors of A such that $Re\lambda_i > 0$.

INDEX